Animal Genetic Engineering

Animal Genetic Engineering: Of Pigs, Oncomice and Men

Edited by

PETER WHEALE and RUTH McNALLY

First published 1995 by Pluto Press
345 Archway Road, London N6 5AA

British Library Cataloguing in Publication Data
A catalogue record for this book is available from the British Library

ISBN 0 7453 0754 X (hbk)

Library of Congress Cataloging in Publication Data
Animal genetic engineering: of pigs, oncomice and men / edited by
Peter Wheale and Ruth McNally.
p. cm.
Based on a conference held at the Royal Geographical Society, London in October 1992.
Includes bibliographical references and indexes.
ISBN 0 7453 0754 X (hb)
1. Animal genetic engineering—Congresses. 2. Livestock—Genetic engineering—Congresses. 3. Animal genetic engineering—Moral and ethical aspects—Congresses. 4. Transgenic animals—Congresses. 5. Animal biotechnology—Congresses.
I. Wheale, Peter. II McNally, Ruth M.
QH442.A55 1995
660'.65—dc20 95–2547
CIP

Typeset from the editors' disks in Iowan Old Style
Designed and produced for Pluto Press by
Chase Production Services, Chipping Norton, OX7 5QR
Printed in the EC by T J Press, Padstow, England

Contents

List of Abbreviations

ACGM Advisory Committee on Genetic Manipulation (HSE, UK)
ACNFP Advisory Committee on Novel Foods and Processes
ACOST Advisory Committee on Science and Technology
ACRE Advisory Committee on Releases to the Environment
ACTIP Animal Cell Technology Industrial Platform
AFRC Agriculture and Food Research Council (UK)
AI artificial insemination
AIDS Acquired immune deficiency syndrome
APC Animal Procedures Committee (HO, UK)
BAP Biotechnology Action Programme (EU)
BBRG Biotechnology Business Research Group
BCC Biotechnology Coordination Committee (EU)
BGH bovine growth hormone
BIA Bio Industry Association (UK)
BLG beta-lactoglobulin
BRIDGE Biotechnology Research for Innovation, Development and Growth in Europe
BSE bovine spongiform encephalopathy
BST bovine somatotropin/somatotrophin
BTG British Technology Group
BUAV British Union for the Abolition of Vivisection
CAP Common Agricultural Policy
CBI Confederation of British Industry
CEC Commission of the European Communities
CEN *Comité Européen de Normalisation* (European Communities' Committee for Standardisation)
CIWF Compassion in World Farming (UK)
CNS central nervous system
COBAF Committee on Biotechnology and Food (UK)
CSIRO Commonwealth Scientific and Industrial Research Organisation
CVMP Committee for Veterinary Medicinal Products (EU)
DES Department of Education and Science (UK)
DNA deoxyribonucleic acid
cDNA complementary DNA
rDNA recombinant DNA
DOE Department of Employment (UK)
DoE Department of the Environment (UK)
DoH Department of Health (UK)
DPAG Dangerous Pathogens Advisory Group (DoH, UK)
DSS Department of Social Security (UK)

DTI Department of Trade and Industry (UK)
EBCG European Biotechnology Coordinating Group
EBIS European Biotechnology Information Service
EC European Community
ECU European Currency Unit
ECVAM European Centre for the Validation of Alternative Methods (to Animal Testing)
EDF European Defense Fund (USA)
EEC European Economic Community
ELISA enzyme-linked immunosorbent assay
EPA Environmental Protection Agency (USA)
EPC European Patent Convention
EPO European Patent Office
ETAAAB European Trade Association for Advanced Animal Breeding
EU European Union
FAC Food Advisory Committee (MAFF, UK)
FAO Food and Agriculture Organisation (UN)
FAWC Farm Animal Welfare Council (MAFF, UK)
FC Food Commission (UK)
FDA Food and Drug Administration (USA)
FDCA Food, Drug and Cosmetic Act (USA)
FIFRA Federal Insecticide, Fungicide and Rodenticide Act (USA)
FoE Friends of the Earth
GAO General Accounting Office (USA)
GATT General Agreement on Tariffs and Trade
GEOs genetically engineered organisms
GF Genetics Forum (UK)
GILSP Good Industrial Large-Scale Practice (EC)
GLSP Good Large-Scale Practice (ACGM/HSE)
GMAs genetically manipulated animals
GMOs genetically manipulated organisms
GMMOs genetically manipulated microorganisms
GNP Gross National Product
HGH human growth hormone
HIV human immunodeficiency virus
HMSO Her Majesty's Stationery Office (UK)
HO Home Office (UK)
HPRT hypoxanthine-guanine phosphoribosyl transferase
HR House of Representatives (USA)
HSC Health and Safety Commission (DOE, UK)
HSE Health and Safety Executive (DOE, UK)
IBA Industrial Biotechnology Association (USA)
IGAP Institute for Grassland and Animal Protection (AFRC, UK)
IGF insulin-like growth factor
IISC Intentional Introduction Sub-Committee

IPR intellectual property rights
IRDAC Industrial Research and Development Advisory Committee (EC)
IVF *in vitro* fertilisation
JPO Japanese Patent Office
LDL low-density lipoprotein
MAFF Ministry of Agriculture, Fisheries and Food (UK)
MECU Million European Currency Units
MEP Member of the European Parliament
mg milligram
MNC multinational corporation
MOD Ministry of Defence (UK)
MOET multiple ovulation and embryo transfer
MRC Medical Research Council (UK)
MRCVS Member of the Royal College of Veterinary Surgeons
NAVS National Anti-Vivisection Society (UK)
NCC Nature Conservancy Council (UK)
NEPA National Environmental Policy Act (USA)
NERC Natural Environment Research Council (UK)
NFU National Farmers' Union (UK)
NGOs non-governmental organisations
NIH National Institutes of Health (USA)
NOAH National Office of Animal Health
OECD Organisation for Economic Co-operation and Development
OSTP Office of Science and Technology Policy (USA)
OTA Office of Technology Assessment (USA)
PAHO Pan American Health Organisation
PR public relations
PRSC Planned Release Sub-Committee (ACGM, UK)
PSF Parents for Safe Food
PTO Patent and Trademark Office (USA)
RAC Recombinant DNA Advisory Committee (NIH, USA)
RCEP Royal Commission on Environmental Pollution
RSPCA Royal Society for the Prevention of Cruelty to Animals
RTD Research and Technological Development
RTDFP3 Research and Technological Development Framework Programme Three (EU)
RTDFP4 Research and Technological Development Framework Programme Four (EU)
SAD Street Alabama Dufferin (attenuated rabies virus)
SAGB Senior Advisory Group on Biotechnology
SEA Single European Act 1986
STOA Scientific and Technological Options Assessment (unit of European Parliament)
TNC Transnational company
TPA tissue plasminogen activator

TSCA Toxic Substances Control Act (USA)
TUC Trades Union Congress (UK)
UK United Kingdom
UN United Nations
UPOV International Union for the Protection of New Varieties of Plants
USA United States of America
USDA United States Department of Agriculture
US PTO United States Patent and Trademark Office
VPC Veterinary Products Committee (UK)
V-RG vaccinia-rabies glycoprotein G (recombinant virus)
WHO World Health Organization

Notes on Contributors

Professor Robin Attfield is Professor of Philosophy in the University of Wales, Cardiff, where he chairs the Philosophy Board of Studies and shares in the work of the Centre for Applied Ethics. His works include, *The Ethics of Environmental Concern and Values, Conflict and the Environment,* a report on environmental decision-making which he edited in 1989 for the Ian Ramsey Centre, Oxford, and *International Justice and the Third World,* which he edited with Barry Wilkins for Routledge in 1992. He co-organised a conference of the Royal Institute of Philosophy in Cardiff in July 1993 on Philosophy and the Natural Environment.

Hiltrud Breyer MEP, is a member of the European Parliament representing the Green Party. She sits on the European Parliament's Committee for Energy, Research and Technology and the Committee for Public Health and Consumer Protection. She has been Rapporteur on the European Parliament's report on Human Genome Analysis and on the European Commission's Communication, *Promoting a Competitive Environment for the Industrial Activities Based on Biotechnology Within the Community* (SEC(91)629 Final).

Professor Derek Burke is the Vice-Chancellor of the University of East Anglia. He trained as a chemist at the University of Birmingham and obtained his PhD from Yale University. From 1955 to 1960 he worked on influenza virus and interferon at the National Institute for Medical Research and then went to Aberdeen University as a lecturer in Biochemistry. In 1969 he was appointed Professor of Biological Sciences at the University of Warwick where he continued his work on molecular biology of viruses and on interferon. He led a group which isolated clones of the human interferon genes and also made the first monoclonal antibody against human interferon. Professor Burke was President of the Society for General Microbiology from 1987 to 1990 and is currently Chairman of the Advisory

Committee on Novel Foods and Processes (ACNFP), a member of the Cancer Research Campaign (CRC) Council and Chairman of the Council of the CRC Paterson Institute.

Joyce D'Silva is a Director of Compassion in World Farming (CIWF) and of Compassion in World Farming Trust (formerly The Athene Trust). She grew up on a farm and obtained her degree at Trinity College, Dublin University. She taught in schools in India and the UK before joining the CIWF in 1985, becoming a director in 1991. Joyce has addressed many and varied audiences on animal welfare issues, including the European Parliament's Intergroup on Animal Welfare and has had articles published in numerous magazines and journals.

Dr Elaine Dzierzak is a Senior Staff Scientist at the National Institute for Medical Research, Mill Hill, London. She was born in Chicago and took degrees at the University of Illinois including a Master of Science in Veterinary Pharmacology. She was awarded her PhD from Yale University for studies of immunoglobulin diversity. She worked as a postdoctoral fellow in the laboratory of Dr Richard Mulligan at the Whitehead Institute, MIT, Cambridge, Massachusetts, where, from 1985 to 1989, she performed studies on retroviral-mediated gene transfer. Her laboratory at the National Institute for Medical Research is presently focusing on an examination of the haematopoietic system during embryonic and foetal stages of development and also with generating a molecular therapy for HIV positive patients.

Dr Ron James, who presented the results of Clarke *et al.* (Chapter 14), is the Managing Director of Pharmaceutical Proteins Limited, Edinburgh. He obtained a degree in Chemistry through part-time study while working as a Laboratory Assistant at Glaxo and his PhD from Imperial College, London. He has spent many years in industrial research and research management with Wilkinson Sword. In 1987 he was involved in the founding of Pharmaceutical Proteins Limited and in 1989 became its Managing Director. The Company is at the forefront of molecular pharming developments.

Dr Tim Lang, is Director of Parents for Safe Food (PSF), a public education and campaigning organisation based in London. He has a PhD in Social Psychology from Leeds University. In 1990 Parents for Safe Food was awarded the prestigious Glenfiddich Special Award for its 5 years of work to improve food quality. From 1982 to 1984 he was Director of the Food Policy Unit at Manchester Polytechnic and Director of the London Food Commission from 1984 to 1990. In 1988 he was awarded an honorary Professorship at Ryerson Polytechnical Institute, Toronto. He is also a founder member of the National Food Alliance and the Public Health Alliance and is on the Board of the *Food Magazine*. He has many published articles on food quality and safety and has co-authored several books including *Food Irradiation: The Facts* (1986), *Food Adulteration and How to Beat It* (1988) and *The Safe Food Handbook* (1990).

Dr Gill Langley is the Scientific Advisor for the Dr Hadwen Trust for Humane Research. She has a Zoology degree and PhD in neurochemistry from Cambridge University and is a Fellow of the Royal Society of Medicine and a Member of the Institute of Biology. She is a proponent of animal rights and since 1979 has worked in the anti-vivisection and humane research movement. She has written books, newspaper articles and papers about animal welfare which have appeared, for example, in the *New Scientist* and the *Journal of Biological Education*.

Dr Maurice Lex, Commission of the European Communities, DG XII, rue de la Loi, 200, B-1049 Brussels, Belgium.

Meredith Lloyd-Evans is a Veterinary Surgeon and Principal Consultant with BioBridge Consultancy, Cambridge. Having qualified as veterinary surgeon in 1973 he practised in Canada and England until 1977 when he joined Smith Kline Animal Health Limited as an expert providing technical assistance for a range of veterinary medicines and products, including vaccines, used to improve farm animal health and productivity. In 1985 he joined the British Technology Group (BTG), where he managed a portfolio of projects, patents and licences in animal health, biomaterials and biotechnology. In 1987 he joined the agricultural biotechnology group within PA Technology and on its dissolution in 1988 founded BioBridge Consultancy. He is a regular contributor to *Animal Pharm* and has written for *Guinness Encyclopedias* and is a member of the BioIndustry Association (BIA) working party on Public Affairs.

Dr Sue Mayer is Director of Science, Greenpeace UK. She has a degree in Pharmacology and a Veterinary degree and a PhD in cell biology/immunology from Bristol University. She has several years of experience in veterinary practice and from 1987 to 1989 she lectured in Veterinary Therapeutics and Toxicology at Bristol University. From 1989 to 1990 she was Director of the RSPCA Seal Unit in Norfolk, leaving there to join Greenpeace UK where she has particular responsibility for genetic engineering.

Ruth McNally has a degree in Genetics from Nottingham University and an MA in Socio-Legal Studies from Brunel University and is presently completing a PhD on the role of the new genetics in society. She has written many articles on genetic engineering and is the co-author of *Genetic Engineering: Catastrophe or Utopia?* (1988) and the co-editor of *The Bio-Revolution: Cornucopia or Pandora's Box?* (1990). She is a Director of Bio-Information (International) Limited, London and, since 1993, also a Visiting Lecturer in the Sociology Department of the University of the West of England.

Dr Ivan Morrison is Head of Immunology and Pathology at the AFRC Institute for Animal Health, Compton, UK. He took his Veterinary degree and PhD at the University of Glasgow. From 1975 to 1989 he worked at the International Laboratory for Research on Animal Diseases focusing his research on the immunology of parasitic infections. Since 1990 he has been Head of the Division of Immunology and Pathology at the AFRC Institute for Animal Health.

Professor Christopher Polge CBE is Director of Animal Biotechnology Cambridge Limited, Animal Research Station, University of Cambridge. He obtained his first degree in Agriculture from the University of Reading. For several years he worked for the Medical Research Council (MRC) and for over 30 years for the Agriculture and Food Research Council (AFRC) in Cambridge. He is a specialist in reproductive biology, low temperature biology and the application of research in reproduction to advanced animal breeding. He pioneered research on the deep freezing of semen which has revolutionised cattle breeding worldwide and later he has been involved in the control of reproduction and embryo transfer. On retirement from the AFRC in 1986 he was co-founder of Animal Biotechnology Cambridge. His work has been recognised by a number of awards, including the Wolf Foundation Prize in Agriculture and the Japan Prize from the Science and Technology Foundation of Japan.

Jeremy Rifkin is Director of the Foundation on Economic Trends, Washington DC, USA. He has an Economics degree from the University of Pennsylvania and an MA in International Affairs from Tufts University. He has campaigned against such technologies as 'factory farming', nuclear power and genetic engineering. He has also testified before numerous Congressional committees and pursued litigation against the government to ensure responsible government policies on a variety of environmental and technological issues. He is author of 12 books, many of which have been translated into foreign languages. For over 25 years his main concern has been to inform and educate the public on environmental issues and science and technology related policies.

Richard Ryder studied experimental psychology at Cambridge University and Columbia (New York) and was for many years a clinical psychologist in Oxford. In 1970 he published *Speciesism* and in so doing created a new word for the welfarist's dictionary. Since then he has written many books and articles on animal welfare. He was Chairperson of the RSPCA Council from 1977 to 1979 and Chairperson of the Liberal Animal Welfare Group from 1980 to 1988. He is currently Director of the Political Animal Lobby Limited in Devon, UK.

Peter Stevenson is Political and Legal Director of Compassion in World Farming (CIWF), Petersfield, UK. He read Economics and Law at Cambridge University and before joining CIWF he worked for the Law Society. He regularly lobbies the European Commission and Members of the European Parliament on animal welfare issues and has drafted several animal welfare Bills for the British Parliament. He has also published highly acclaimed reports on the slaughter of pigs, cattle, sheep and chickens.

Dr Kevin Ward works for the CSIRO Division of Animal Production, where he is Manager of the Wool Biotechnology Programme, based in Blacktown, Australia. He obtained his first degree in Biochemistry from the University of Sydney and his PhD from the University of Massachusetts. He has been with the CSIRO Division of Animal Production since 1973 where he leads a programme of research into a range of topics associated with the improvement of wool growth, including the genetic engineering of sheep. More recently, he has worked in the fields of disease-resistance of sheep and cattle.

Professor John Webster is Professor of Animal Husbandry in the School of Veterinary Science, University of Bristol. He has studied the nutritional and environmental requirements for health and production in cattle for over 25 years. He has also conducted research into the welfare of veal calves – the Access System – in which the social, physiological and nutritional needs of the animals are fully met. He is a member of the Farm Animal Welfare Council (FAWC) and has written many articles on animal welfare and *Understanding the Dairy Cow* (1987).

Dr Peter Wheale has a degree in Economics from Manchester University and was awarded his doctorate for studies on the conduct, structure and performance of the cereals industry. He also has an MSc in the Structure and Organisation of Science and Technology from Manchester University and an MA in Medical Ethics and Law from King's College, London. He has written articles on science and technology policy, economics and bioethics. He is co-author of *People, Science and Technology: A Guide to Advanced Industrial Society* (1986), *Genetic Engineering: Catastrophe or Utopia?* (1988) and the co-editor of *The Bio-Revolution: Cornucopia or Pandora's Box?* (1990). He currently holds a Senior Lectureship at the European Business School, London and is also Chairperson of the Biotechnology Business Research Group (BBRG) and a Director of Bio-Information (International) Limited, London.

Dr [illegible] is [illegible] Director [illegible] [illegible] [illegible]

[illegible] John Webster is Professor of Animal Husbandry, [illegible] School of Veterinary Science, University of Bristol. He [illegible] [illegible] [illegible] and [illegible] [illegible] [illegible] of animals [illegible] and [illegible] [illegible] [illegible] has written [illegible] Animal Welfare [illegible]

Dr Peter Wheale [illegible] [illegible] and [illegible] [illegible] [illegible] and [illegible] [illegible] Genetic Engineering: Catastrophe or Utopia? (1988) and [illegible] [illegible] [illegible] [illegible] and [illegible] [illegible]

Preface

Animal Genetic Engineering: Of Pigs, Oncomice and Men is based on the highly acclaimed Conference which took place in London at the Royal Geographical Society in October 1992. The Conference was organised by the educational wing of Compassion in World Farming (CIWF), which is dedicated to the promotion of harmony between people and the natural world. Twenty contributors, including animal welfare campaigners, scientists, civil servants, bio-industrialists and philosophers, each of them experts in their respective fields, explain and debate the controversial issues surrounding the genetic engineering of animals.

The topics covered in this volume include debates on the utility of producing transgenic farm animals, embryo transfer, cloning and reproduction, genetically engineered veterinary vaccines, the production of therapeutic proteins using transgenic animals and the animal welfare implications of these various developments. The issues surrounding the patenting of transgenic animals are also explored, as are the ethics of animal experimentation and the morality of killing animals to harvest their organs for human transplantation. Several contributors consider the environmental hazards and public policy issues posed by transgenic manipulation technologies, and the political issues raised by our attempt to regulate these various activities are vigorously debated.

An introductory chapter is provided by Richard Ryder which briefly outlines each contributor's focus of interest and introduces the reader to many of the important controversial issues posed by the genetic engineering of animals. The following five parts of the volume are respectively entitled: Farm animal biotechnologies; Transgenic farm animals: Regulation and impact; Patenting of genetically engineered animals; Genetic engineering of laboratory animals; and What place has genetic engineering of animals in society? Each part of the volume includes an edited transcription of the lively open forum discussions which took place at the Conference itself.

As editors, we have, where we thought it appropriate, inserted notes and updated certain contributor's statements particularly where important developments have since occurred. Such editors' inserts are clearly marked. We have also provided the reader with a comprehensive glossary of technical terms, a list of abbreviations, name and subject indexes and a comprehensive bibliography.

We believe the volume will be useful to a wide range of readers, including humanities and social studies students and life science students who wish to gain a better understanding of the social relations of this field of science and technology and the ethical dilemmas posed by the genetic engineering of animals. We should emphasise, however, that no specialised knowledge is required to understand the ideas and arguments presented by the respective contributors.

Peter Wheale and Ruth McNally
London, March 1995

1 Animal Genetic Engineering and Human Progress: An Overview

Richard Ryder

In this opening chapter the main topics presented in *Animal Genetic Engineering: Of Pigs, Oncomice and Men* are introduced. The author's own views on genetic engineering and animal welfare are also expressed.

FARM ANIMAL BIOTECHNOLOGIES

The chapter by environmental activist Jeremy Rifkin, Director of the Foundation on Economic Trends, USA, is based on his opening address to the Compassion in World Farming international conference on animal genetic engineering which forms the basis of this book. Rifkin begins by questioning whether or not we, as a society, really need animal biotechnology at all. He is concerned at the way in which genetic engineering is being presented as a 'quick fix' for the world crises of desertification, ozone depletion and global warming, and asserts that the 'real' solution, that is to achieve 'sustainability', is not being tackled seriously.

Apart from questioning whether it is morally right to transfer genes from species to species, Rifkin also asks who should regulate biotechnology: should it be regulated by governments, industry or consumer groups? He predicts that, despite assurances to the contrary, many of the environmental releases of genetically manipulated (or genetically engineered) organisms (GMOs) that are being undertaken will not be safe because the technology for risk-assessment is simply not sophisticated enough to provide accurate information on how GMOs will behave in the environment. The fact that companies wanting to release GMOs cannot obtain insurance against the possibility of biological pollution resulting from a deliberate release alerts us to the perceived risk. Such pollution could be worldwide, since once released, GMOs cannnot be contained within national boundaries. Rifkin argues that the fact that the capitalist market-place will not provide insurance against risks of biotechnology should be a signal to us that no releases should be taking place. His position is that any human intervention in the environment should be conservative, cautious and elegant, and that the deliberate release of GMOs cannot be described in these terms.

Rifkin goes on to consider one of the first major biotechnology products for agriculture, bovine somatotropin (BST), a genetically engineered hormone which is claimed to increase milk yield in dairy cows. Over US$1 billion has been spent on developing this product for which, Rifkin contends, the industrial world has no use. The European Community (EC) and the USA, which are awash with milk already, do not need a product to boost milk production. Taxpayer's money is already being spent to buy up surplus milk from farmers; surely it is illogical to subsidise even higher dairy production.

Turning to safety issues, Rifkin alleges that companies involved in developing and testing BST have not declared adverse results on animal safety which indicate that the widespread use of BST will lead to an increased incidence of mastitis amongst dairy cows, the treatment of which could lead to higher levels of antibiotic residues in milk. He expresses concern about the safety of milk produced with the aid of BST for human consumption. Despite pressure from the bio-industries for the commercialisation of BST, Rifkin asserts that it is not viable as a commercial product and declares that the campaign against its use in dairy farming will continue. Rifkin hopes that this will be the first of many consumer-led anti-biotechnology campaigns that will force more public debate over biotechnology before products come on-line.

Rifkin also considers the controversy surrounding animal patenting. One problem he identifies is that of control of the gene pool by those who own the patents on life forms. This is a particular problem because the gene pool resides in the less developed South, yet the technology to control and manipulate it is in the North. Rifkin suggests that at the Rio Summit in 1992, the Northern multinational companies opposed the biodiversity treaty because of their concern over patenting and property rights. This has led to developing countries closing their gene pools to the North. Rifkin argued that both these extreme stances are misguided and suggests that although the gene pools should not be commercialised they must remain open to all.

Rifkin presents a radical alternative to looking at our use of resources. He contends that if we regard progress in terms of generating more production, 'silly' products like BST are the result. He proposes that the quest for efficiency should be replaced by a move towards sufficiency. He is particularly sceptical of the role of animal biotechnology and animal patents within such a philosophy.

In his conclusion, Rifkin mentions two of his ongoing campaigns. First, the Pure Food movement in the USA which is fighting to prevent any genetically engineered foods reaching the market-place. This campaign has recruited chefs to display a symbol in their restaurants which indicates that no genetically engineered foods will be served on the premises. Rifkin is aiming for a permanent, irreversible boycott of genetically engineered foods. He anticipates a huge consumer resistance to such foods, dubbed 'frankenfoods', and suggests that the introduction of these products goes against an established consumer preference for organic foods. He also wishes to see clear labelling of genetically engineered foods and is cynical about the industry's reluctance to support such moves. He asserts that the public has a right to know what genes have been inserted into new products, particularly as the origin of the introduced genes could contravene religious and moral codes. He claims that industry knows that if it has to label new foods as the product of genetic engineering, the public will not buy them. He asks, for example, would a vegetarian want to eat crops in which an animal gene had been inserted, or a Jew want to eat fish or beef in which a pig gene was expressed? The second new campaign, 'Beyond Beef', aims at reducing the North's consumption of beef products. The campaign will encourage people to reduce their consumption of beef by 50 per cent and persuade them to take a step down the food chain.

Dr Kevin Ward's contribution includes consideration of some of the most contested population statistics. Ward is the Manager of the Wool Biotechnology Programme, CSIRO, Australia. He claims that while the human population of the planet is increasing exponentially, food production is only edging upwards slowly. He argues that therefore we have to choose between curbing the human population, limiting standards of living or enhancing food production. Ward's own research focuses on transgenic sheep designed to produce up to twice as much wool as normal sheep. If this experiment proves technically and commercially viable Ward believes that the Australian sheep population could be halved with no losses in wool production, thereby reducing the demands of sheep farmers on land and water resources for their herds.

Ward discusses research aimed at introducing the plant gene chitinase into sheep as a form of insecticidal protein to protect the animals from fly-strike. The idea here is that insect larvae grazing on the sheep would die from ingesting the insecticidal protein before injuring the animal. Ward points out the environmental benefits of having an indigenous insecticide which would lead to a reduction in chemical spraying. Furthermore, Ward argues that if the incidence of fly-strike were reduced, substantial benefits could be gained in animal welfare and health. Ward's assertions are questioned from the floor (see Discussion I), but he strongly defends his thesis that judicious use of genetic engineering technology 'may hold part of the key to the long-term survival of human civilisation'.

Dr Ivan Morrison who is the Head of Immunology at the Agriculture and Food Research Council (AFRC), Institute for Animal Health in Compton, gives examples of how genetic engineering and associated technologies may produce benefits to people, for example, through the production of vaccines and in the cheaper production of food animals. Morrison argues that a major cause of animal suffering is disease which could be reduced through genetically engineered vaccines and transgenic animals with disease-resistance genes.

Morrison attempts to counter critics (see Discussion I) who suggest that improved disease-resistance would result in even more 'factory farming' by suggesting that many farmers keep more animals than they need because they know that there will be a high mortality due to disease. Pressure on land and food resources would be reduced by the use of disease-resistant transgenic animals as the number of animals required to give the same final yield could be reduced. However, Morrison is realistic about the technical difficulties which face scientists attempting to produce disease-resistant transgenic farm animals. There is a lack of knowledge of the genetic characteristics of many animal diseases, and where such diseases have multi-factorial causes Morrison concedes it may prove impracticable to use genetic engineering.

Professor Christopher Polge, the Director of Animal Biotechnology Cambridge Ltd, focuses on artificial breeding techniques and the improvements that have been achieved in animal productivity. Embryo transfer and the ability to distribute desirable cattle semen to herds on a global scale has, he believes, revolutionised animal breeding programmes. He argues that where indigenous breeds are being replaced by animals bred from imported semen, the indigenous gene pool could be cryopreserved through the freezing of semen.

Polge describes the techniques of *in vitro* fertilisation (IVF), pre-selection of embryo sex, the splitting of embryos and nuclear transfer. He also refers to legislation designed to regulate the use of these techniques and protect animal welfare. He believes genetic engineering offers a potential for alternatives to *in vivo* testing (vivisection) and for the preservation of rare species.

TRANSGENIC FARM ANIMALS: REGULATION AND IMPACT

The first chapter in this part of the book is by John Webster, Professor of Animal Husbandry at Bristol University, who discusses the impact of genetic engineering on animal welfare. He begins by stating that animals are free from introspection and thus are unlikely to worry about whether they are patentable or not. I agree with his contention that the motivation behind animal experiments is not important to the animal; from the animal's perspective, only the effects that the experiments have on its quality of life matter. Webster argues that this should be borne in mind when such experiments are contemplated, and this leads him to ask: were it to become possible to engineer animals incapable of experiencing pain or distress would it be morally justifiable to do so? To this question I, unlike Webster, am tempted to reply that it might even be our moral *duty* to do so! I strongly suspect that breeding creatures who are made positively happy by being exploited may be a good thing, and I find talk about lack of dignity, sacredness of life, *telos*, respect for the biotic community, and so on, rather unconvincing as a basis for ethics. I believe suffering to be the only reliable basis for morality, and that genetic engineering is only wrong if it causes suffering to 'painient' beings (beings capable of experiencing pain, see also below).

Webster outlines seven areas of genetic engineering likely to affect animal welfare and assesses their probable impact. He suspects that predictions, such as those made by Ward that wool production might be doubled by the insertion of bacterial genes into sheep, are over-optimistic, and he suggests that enthusiasm for such applications should be curbed. Webster contends that the manipulation of body size, shape or composition was already out of favour with good biotechnologists. The problems experienced with transgenic animals carrying extra copies of growth hormone genes, the multifactorial nature of such traits, and the controversy over the use of BST have all led to a more cautious approach to research in this field.

Webster suggests that laboratory animals are better protected under British regulations than are farm animals since under the Animals (Scientific Procedures) Act 1986 applications to perform experiments with laboratory animals must include a form of welfare cost–benefit analysis. He concludes that the same sort of law should be applied to farm animals in order to protect the farmer, animals and the public from some of the less acceptable applications of biotechnology. I have to say that I believe such cost–benefit exercises may be good administration but make poor ethics. In my view, the ends can never justify the means when the means constitute cruelty.

Professor Derek Burke is Chairman of the Advisory Committee on Novel Foods and Processes (ACNFP) of the Ministry of Agriculture, Fisheries and Food (MAFF). He describes the work of this Committee, which is one of the many official committees now required to evaluate the impact of genetic engineering, in this instance chiefly from the perspective of consumer safety. Personally, I am a little puzzled by this Committee's composition – why two ecclesiastics, for example?

Burke considers that three classes of transgenic animals are likely candidates for human consumption. First, transgenic animals which are actually expressing a foreign gene; secondly, transgenic animals in which the transgene has integrated but where no foreign protein production is detectable; and thirdly, transgenic animals that have been microinjected with foreign DNA but do not show any evidence of DNA integration. He suggests that there should be no problems in eating any of these categories of transgenic animals since we consume foreign DNA each time we pick up a new exotic fruit from the supermarket. However, he accepts that the sale of transgenic animals for human consumption is likely to continue to be of concern to the general public.

Dr Tim Lang is Director of Parents for Safe Food. He is critical of the composition of official committees such as the ACNFP. He suggests that the more consumers know about genetically engineered food, the more worried they become. Lang is highly critical of the attitudes of large multinational food companies to consumer interests and preferences on food labelling policies. He quotes the results of a 1991 UK survey which suggested that 25 per cent of the public would want genetically engineered food to be labelled as such. A similar survey in 1992 revealed that 85 per cent wanted labelling but that 48 per cent of consumers would consume products derived from genetic engineering. He cannot understand why food companies are so against labelling when it is clear that at least a proportion of the public are willing to buy such foods. He is worried by the possibility that the Food and Drug Administration (FDA) in the USA may allow 'nature identical' products to reach the market without adequate labelling. Lang supports Rifkin's campaign to have all genetically engineered food products labelled. He suggests that liberal new EC regulations on genetically engineered food products are the result of pressure exerted by multinational corporations and alleges that when the European Commission tried to introduce tougher regulations on genetically engineered foods, companies threatened to withdraw their activities from Europe taking the jobs they had created with them.

In conclusion, Lang proposes that the British government should set up an ethics committee to deal with genetic engineering and biotechnology and that a moratorium on the development of food products should be called to allow time for public education and consultation. He also stresses the need for worldwide alliances between non-governmental organisations (NGOs) concerned about animal patents and consumer safety if effective opposition to genetically engineered foods is to be successful.

Joyce D'Silva is the Director of the educational wing of

CIWF and one of the organisers of the Conference. She brings us back to the central issue of *Animal Genetic Engineering: of Pigs, Oncomice and Men* – the welfare of animals. She presents a critical view of the genetic engineering of farm animals, reminding us that animals are being treated as mere production machines. Modern broilers, turkeys, pigs and cows already suffer painful conditions due to traditional breeding methods. She asks, 'How much more may be the suffering of genetically engineered creatures?'

D'Silva counters the claim that genetic engineering is accurate and predictable by quoting from many transgenic animal studies in which foreign gene expression has produced unanticipated adverse effects. For example, transgenic lambs suffer from diabetes, and transgenic pigs from arthritis and gastric ulcers. She argues that genetically identical animals are vulnerable to disease, and is sceptical about claims that increasing disease-resistance in animals through genetic engineering will improve animal welfare, and considers it to be a poor substitute for improving their living conditions. D'Silva concludes by recommending that we should change our perception of animals as reproductive machines to one of seeing them as sentient creatures capable of suffering and worthy of our respect.

Ruth McNally is a researcher at the University of the West of England in Bristol, and a director of Bio-Information (International) Limited. She examines one case where success has been claimed – the genetically engineered rabies vaccine – and concludes that even here there are risks of affecting the environment which are disturbingly difficult to assess.

McNally focuses on the release of recombinant rabies vaccine in Europe. Already, a recombinant vaccine based on the vaccinia virus has been used in France and Belgium as part of a fox rabies eradication programme. The programme involves large-scale releases of a genetically engineered virus. She considers the safety aspects and the practicality of the fox rabies eradication programme, and argues that the aim of the programme – to have a rabies-free Europe is unrealisable and its effects hazardous.

The problem, McNally asserts, lies in the population dynamics of the main rabies virus reservoir – the red fox. Rural areas have very high fox populations, and it has been calculated that to control the spread of rabies, 80 per cent of the fox population would have to be immunised. Since foxes produce 4–5 cubs per year and only have a life expectancy of 1.5 to 2 years, it is very difficult to maintain high levels of herd immunity in the fox population. She is concerned that the vaccinia virus could recombine with other closely related viruses in the environment resulting in new viruses, for example, a new form of the pox virus.

In the final chapter in this part of the book, Dr Sue Mayer, Director of Science for Greenpeace, UK, further expounds upon the theme of the safety issues surrounding the release of GMOs. She argues that it is the disruption of cross-species barriers made possible by modern genetic engineering which makes it potentially so dangerous. She reminds us that once released into the environment there may be no way to recall GMOs.

Mayer expresses her concern about the inadequacy of risk-assessment procedures. Most releases so far have involved microorganisms and plants, organisms that are more difficult to control than animals. For example, no economically feasible method has yet been devised to contain pollen from genetically engineered plants, so there is a real danger of their cross-pollinating with wild relatives.

Mayer goes on to compare the release of genetically engineered fish with the escape of farmed fish. Escapees from fish farms, she claims, have carried disease into the natural fish populations of Scotland and Norway and have changed the genetic basis of the natural populations by interbreeding. It has been suggested that certain transgenic fish could be engineered to express a foreign 'antifreeze gene' to enable their geographical range to be extended to colder climates. Mayer contends that the effect of such a transgenic animal on the ecosystem is impossible to predict.

PATENTING OF GENETICALLY ENGINEERED ANIMALS

In the opening chapter of this part of the book, Meredith Lloyd-Evans, a consultant with BioBridge UK, considers the issues surrounding the patenting of genetically engineered animals. He argues that patents are essential if the developers of an invention are to recoup their costs.

Lloyd-Evans asserts that most transgenic animal patent

applications relate to disease models and to the production of pharmaceutical proteins. He describes what he considers to be acceptable and unacceptable uses of transgenic technology. He predicts that any patents in which the characteristics of farm animals were drastically changed, for example, a wingless chicken or a cow with a second udder, would be refused. Similarly, he thinks the development of new companion animals in different shapes and colours would be unethical. However, he points out that in Asia conventional breeding of novel decorative fish is popular, and asks should the means of producing novel fish rather than the resulting phenotype determine whether the animal is morally acceptable?

In his conclusions, Lloyd-Evans suggests that industry should set up its own ethical frameworks in conjunction with consumer groups to determine what types of animals are produced and patented. It would, he suggests, be in the bio-industries' own interests to support animal welfare legislation and work with animal welfare groups. In his opinion, banning patenting would not benefit animals or mankind.

In direct contrast to Meredith Lloyd-Evans, Peter Stevenson, a lawyer working for CIWF, presents the case for banning animal patenting. Stevenson considers the types of applications likely to come before the European Patent Office (EPO) which include chickens carrying a cattle growth hormone gene. Is such an invention necessary, he asks, when chickens produced by conventional breeding to be heavier and grow faster already have considerable health problems? As for disease models, Stevenson contends that no animal should be created to suffer despite its potential usefulness to man. He argues that animal welfare values need to be respected even if adhering to them is to human detriment and he emphasises that the European Patent Convention (EPC) does allow patents to be refused on inventions which are contrary to public order or morality.

GENETIC ENGINEERING OF LABORATORY ANIMALS

The first chapter in this part of the volume by Clark *et al.* was presented at the Conference by Dr Ron James, the Managing Director of Pharmaceutical Proteins Limited, Edinburgh. The Edinburgh team of scientists are developing transgenic sheep for the production of antitrypsin in milk. Current methods of producing antitrypsin involve extracting it from human blood which is expensive and inefficient, and at present there is a very limited supply of the drug available. They suggest that producing the protein in the milk of transgenic sheep would be cheaper and more efficient than in transgenic bacteria, and that in this way sufficient quantities to treat all patients could be harvested. In order to target expression to the mammary gland, they have used regulatory sequences from a gene encoding the sheep whey protein beta-lactoglobulin (BLG). The authors suggest that the production of pharmaceuticals by this route is one of the uses of transgenic animals which can be justified on ethical grounds.

Dr Elaine Dzierzak works at the National Institute for Medical Research at Mill Hill, London. She describes certain scientific studies which she suggests indicate how useful transgenic animals can be in the treatment of genetic diseases, the testing of therapies, and as disease models. The species of animals currently most favoured for experimental purposes are rats, mice, rabbits, guinea pigs and rhesus monkeys. All licensed research falls under the Animals (Scientific Procedures) Act 1986.

Dzierzak suggests that the development of transgenic models complements the use of naturally occurring animal models. For example, she asserts that the sickle-cell mouse enabled the testing of new anti-sickling agents and helped scientific research into what precipitates a sickling crisis, and that the development of certain gene therapy techniques has been made possible using transgenic mice. Dzierzak contends that such animal models are very useful as they aid the study of disease and the development of new medical therapies.

In contrast to the views expressed by Dzierzak, Dr Gill Langley of the Hadwen Trust for Humane Research reveals what she believes are numerous telling examples of the unacceptable suffering of laboratory animals subjected to genetic engineering experiments. She describes research which must have caused extreme suffering in the animal subjects. She questions how valid animal models of human disease are when examined closely since animal models do not necessarily produce illnesses comparable to human illnesses. Finally she emphasises, as does Joyce D'Silva, the unpredictability of the consequences of transgenic engineering, pointing out that it can produce completely unexpected results. She cites several cases of unpredictability where instead of being a precise, targeted technology, as its proponents often claim, genetic engineering has had dramatic and unexpected effects that caused the animal subjects to suffer.

WHAT PLACE HAS GENETIC ENGINEERING OF ANIMALS IN SOCIETY?

Robin Attfield is Professor at the Centre for Applied Ethics at Cardiff University. He considers whether unnatural kinds of animals could be morally wronged. He contemplates whether merely bringing a creature into existence could harm it. Were such a creature only to know a life of misery then that would be wrong because it would be a life not worth living. He thus argues that agents do have a responsibility for the quality of the lives they bring about.

Attfield considers the morality of genetically engineering animals that are incapable of feeling pain and suffering so that they would be better suited to 'factory' farming since they would be produced specifically for that purpose. In Attfield's view we should have sufficient respect for other members of the biotic community to not want to change them. However, he could envisage a scenario where just enough of the familiar old animal species were preserved to maintain human sanity!

Attfield suggests that if pain and suffering were the only evils to be considered then it would be morally acceptable to produce animals that are not sentient or which are kept permanently anaesthetised. Personally, I believe that pain and suffering *are* the only evils. In my view, other commonly held evils, such as injustice, loss of liberty or damage to 'the integrity, stability and beauty of the biotic community' (in Aldo Leopold's (1949) phrase), are only evil in as much as they cause pain. However, one should certainly count the annihilation of pleasure as a pain (or an evil) of a sort. A permanently anaesthetised animal (much as a dead one) does not suffer but may be missing pleasures. Does this matter if the animal does not know that s/he is missing out on them?

On the ethical premise that what causes suffering is wrong – unless it benefits that same individual or is undergone voluntarily – then it must be concluded that genetic engineering is not to be automatically condemned as a matter of principle. War, torture or starvation, for example, always cause suffering. But genetic engineering, surely, is not quite in this category. It may cause human or non-human suffering or it may not. Research may be possible without causing suffering as in the case of research on 'non-painient' organisms. It may even, conceivably, sometimes enhance the quality of life and cause happiness.

Dr Peter Wheale is a director of Bio-Information (International) Limited and senior lecturer at the European Business School, London. He considers the ethical aspects of using animals for the benefit of human health. He explains that human kidneys, livers and hearts for transplantation are all in very short supply, and he asks, 'Do animal donors suffer, and should we kill animals in order to harvest their organs for transplanting into humans – the practice called "xenografting"?'

Wheale contrasts two ethical stand points, namely, utilitarianism and contractualism, and discusses how these theories could help us in recognising our responsibility towards animals. He attacks utilitarianism as a doctrine of practical ethics. According to the utilitarian view, animal suffering can always be justified because of the benefit derived by humans, and in this ethical framework, animals invariably matter less than humans. He also rejects the contractualist model, as proposed by John Rawls, in which only rational autonomous agents seem logically to qualify as moral agents.

Wheale, however, argues for a logical extension of Rawls's contractualist model to include certain categories of non-human animal. His argument is based on the idea of a society in which the needs and rights of less able individuals are taken into account. He suggests that advocates could ensure that certain categories of non-human animals be protected in the same way as non-competent humans as, for example, severely mentally handicapped people are. Under such a framework, any animal suffering would have to be justified on culturally acceptable grounds. In the case of xenografting, this would mean that a baboon would be considered as an individual in its own right, and not simply as an organ donor. In his conclusion, Wheale predicts that bioethics will play a bigger role in our future society and suggests that it is already becoming more publicly acceptable to award rights to certain non-human life forms.

My own view is that both rights and duties are human constructs and two sides of the same coin. I believe that all 'painient' things (including aliens, robots and babies) should be given rights; morality should be based upon an individual's 'painience'. Racism, sexism and 'speciesism' overlook 'painience' and are blinded by skin colour, secondary sexual features, quadrupedality and other irrelevant physical appearances. Why do I use 'painience' instead of sentience? Well, sentience covers the capacity to feel all sorts of things besides pain. In principle, a sentient creature might be incapable of feeling pain at all. So we have to find some new words to describe the capacity to feel pain and by pain, of course, I refer to all negative feelings, all sorts of suffering.

Dr Maurice Lex is Scientific Officer to Directorate-General (DG) XII of the Commission of the European Communities. No less than 15 DGs have an interest in genetic engineering. Lex describes the Commission's support for genetic engineering research including the range of research projects which have received EC funding. He asserts that ethical assessment is becoming a more important aspect of research proposals, and claims that 3 per cent of the research budget of relevant parts of the Community's Framework Programme is earmarked for research on the ethical and social effects of biotechnology and environmental risks.

Lex characterises the Commission as a 'listening' body sensitive to public opinion. He summarises the Commission's policy on the release of genetically engineered organisms (GEOs) and maintains that the input of welfare, consumer and environmental groups and industry was taken into account when legislation was being formulated. However, Lex's view of the Commission as an organisation willing to listen to public pressure groups and respond to public opinion is contested by the next contributor, Hiltrud Breyer.

Breyer is a German Green Party member of the European Parliament. She argues forcefully that the controlling influence over EC policy is that of industry and that industry's concern is not with the real risks of genetic engineering but purely with its image in the public's eye. Why, Breyer asks, are so many EC decisions taken behind closed doors? And she expresses her concern that ethical questions are being dealt with by 'closed committees', which public pressure groups are unable to lobby.

Breyer sides with those like Stevenson who are against

animal patenting on the grounds that it would give industry control of the world's germplasm. She asserts that arguments that biotechnology will increase biodiversity are absurd; on the contrary she suggests that it is far more likely to limit the range of animals and plants in common use, ultimately leading to a loss in biological and genetic diversity.

Breyer concludes that a *laissez-faire* attitude to genetic engineering will result in the privatisation of nature and produce generations of animals that will suffer as a consequence. She attacks what she perceives as the lack of controls in the EC and echoes Rifkin's question, that if industry considers deliberate release to be safe why will it not accept, in the form of insurance cover, responsibility for its possible harmful consequences? Taking up the same argument as Mayer, she says that it is most important to get adequate regulatory safeguards in place concerning the release of GMOs as there can be no second chance: a GMO which has become a pest cannot be recalled.

Breyer is not entirely pessimistic about the possibilities provided by the democratic process. After all, animal welfare groups have been able to assert some constraining power over industry as witnessed by the renewal of the European moratorium on BST. She believes that similar results could be achieved in the contemporary political debates on animal patenting and the release of GMOs into the environment.

The important political question is what restraints do we need to keep genetic engineering under control? In Britain the government at least now provides annual figures on the numbers of genetic research projects using animals which is over 200,000 annually. But the mechanisms of control remain inadequate. For example, animal welfare is still not represented on the appropriate committee of the Health and Safety Executive (HSE). In my view, at the very least a named scientist in each project should be held legally responsible for any adverse environmental consequences of the release of transgenic creatures.

Genetic engineering is surely too dangerous to be left to parties with vested interests. It needs a unified and international machinery for inspection, licensing and control. I believe the place of genetic engineering in society is, and should remain, a controversial one, and one constantly out in the open and under scrutiny. One view common to all contributors is that the public must be let in on the decision-making. These are matters for public discussion, for parliamentary debate and the involvement of the consumer, environmentalist, animal welfarist and taxpayer. Governments must recognise that genetic engineering could become a matter of grave importance to the future of the planet. As with nuclear technology, genetic technology may open a Pandora's box. To quote Robert Burns:

> The best laid schemes o'Mice an' Men,
> Gang aft agley,
> An' lea us nought but grief an' pain,
> For promis'd joy!
>
> (*Ode To a Mouse*)

PART I

Farm Animal Biotechnologies

2 Farm Animals and the Biotechnology Revolution: A Philosophical Overview

Dr Jeremy Rifkin

GLOBAL COLONISATION

Five hundred years ago in Tudor England, medieval agriculture was formed collectively around commerce. It survived for six centuries and provided sufficient food for the population of the time. As textile markets developed and new trade centres demanded an increase in sheep production, an important political question emerged concerning animal husbandry and agriculture. The question was, 'Do you raise sheep on the land or do you grow food for people?'

The aristocracy, with a little bit of help from the Church and the newly emergent merchant class, decided to sheep-farm, and this began the radical venture to enclose the commons – to reduce it to a commodified status that could be negotiated as private property in the market-place. Systematically, as readers will recall from their English history, the peasants were moved off the commons. They became a disenfranchised labour force – a development which played a major role in the process of industrialisation. Then began a rare, interesting and novel experiment in human history: the colonial 'error'. In the next five centuries, western European powers began systematically to enclose and commodify – turn to commercial and political property – all the great common land. It has now reached the point where virtually every square foot of land is put to commercial use.

Not satisfied with commodifying and enclosing the common land, we began exploiting the oceans in order to commodify them. Nowadays, under the Law of the Seas Treaty, countries control up to 200 miles beyond their domestic waters, which is about 90 per cent of everything worth owning in the ocean. Having enclosed and commodified the land commons and the great oceanic commons, we next colonised the air commons. We have turned the Heavens, formerly the home of Gods, into air routes which can be leased for the appropriate market price. Having enclosed and commodified the land commons, the oceanic commons and the air commons, we are now after the electromagnetic spectrum. No one knows how we are going to lease that! Finally, this decade, the *fin de siècle,* we have made the radical decision to intervene in the most intimate commons of all – the biological commons.

What has been the consequence of five centuries of enclosing the commons? Global warming, ozone depletion and loss of biodiversity. With all its imperial majesty, the effects of a great civilisation like Ancient Rome, which eroded the land, deforested, and finally collapsed, were relatively local. By contrast, global warming, ozone depletion and loss of biodiversity are a new genre of threat because they are global in scale. We human beings have altered the very biochemistry of the planet.

I do not want to be glib about something so profound. If you measure human accomplishments to date in terms of sheer magnitude, these are the greatest accomplishments of the human race: global warming, ozone depletion, mass deforestation of the equatorial belt, species extinction, desertification and human hunger on a global scale. Global warming is part of the price to be paid for the 'age of progress'. The ozone hole is getting bigger. We are now losing a species to extinction every 60 seconds; we are going to lose 15 per cent of all the living creatures on this planet in the next 8 years. The desertification of our great grasslands is moving at an exponential pace. Twenty per cent of our species will go to bed hungry tonight.

This has been the age of progress only for Europeans, Americans and Japanese. To believe otherwise is either naive or disingenuous. Never before in our history has a fifth of our species gone to bed hungry, systematically, year in year out – not in paleolithic times, not in neolithic times, not in antiquity, not in medieval times.

Sometime in the 1990s we are going to be forced into understanding the transition into a new age which is presently taking place. The fact is, right now, as we go about our business in our various occupations, a new socio-economic global structure is being created. We are moving out of an industrial age based on fossil fuels and into a biotechnological age based on biology and gene pools. This transition will be every bit as important as the transition from medieval agriculture to the Industrial Revolution. It may even exceed it in importance, and it will force us to ask all the basic questions once again.

There are two broad approaches available to us and they differ considerably in philosophy, design, conception and commercial application. One is genetic engineering. The other is a new and sophisticated ecological approach – sustainability, i.e. sustainable relationships and technology. I believe that the choice facing us is between these two approaches.

GENETIC ENGINEERING

Genetic engineering is engineering applied to genetics. It takes engineering principles that we effectively used against inanimate material in the industrial era, and applies them to the genetic code – the blueprint, the architectural design – of microorganisms, plants, animals and humans.

What are these engineering principles? They are concerned with 'quality control' – the ability to predict outcomes with a measure of certainty: quantifiable standards of analysis, utility and efficiency. And now we are talking about taking these principles and applying them to a blueprint of evolutionary design.

When we began to discuss genetic engineering in the 1970s I used to ask audiences this question. Who are we going to trust with the ultimate authority over the blueprints of life? Who decides what are the good and bad genes? Who orchestrates the future evolutionary development on this planet? How many of you would entrust Parliament? The Prime Minister? The British Labour Party? How many of you would entrust multinational corporations (MNCs) to have the wisdom – the clairvoyance – to design the architectural blueprint for millennia of history to come? How many of you would entrust the consumers in the market-place? Who would trust themselves? I have been asking this question for a long time. I have moved through middle age and am now an older man, yet every time I ask this question I never see any hands raised. Should we embark on this journey of reformation when there is no institution or group of individuals that we would be willing to entrust with authority over the genetic blueprints of life? But that is exactly the project on which we are embarked today.

Let us take a look at genetic engineering in agriculture and animal husbandry. Our scientific community and our corporations are now experimenting with releasing genetically engineered microorganisms, plants and animals on a huge scale. Scores of genetically engineered introductions will take place over the next few years, resulting in the release of genetically manipulated organisms (GMOs) in their hundreds, and in their thousands – microorganisms, plants and animals – in massive volumes, released into ecosystems all over the planet. What will be the likely environmental consequences of this global process?

The genetic engineering industry speaks two different languages. When they are talking to investment clubs they say, 'This is the most powerful technology since fire.' When they talk to the public they say, 'This is just an extension of classical breeding.' Is it just an extension of classical breeding? Since the neolithic revolution, we have effectively been crossing some biological boundaries (because taxonomy is arbitrary). We can cross certain plant lines, we can cross a donkey with a horse and get a mule, but we cannot cross a donkey with a dandelion and get a damn thing! Now we are talking about a technology that allows us to combine genes from totally unrelated species, across the entire plant and animal kingdoms, in order to make novelties designed for market performance, and introduce those novelties in massive volumes all over the Earth.

During the Colonial era Europeans introduced organisms – by accident or design – from native to non-native systems. Such organisms either survived by finding a niche or died out. A few of them became very prominent pests to humans, like rabbits in Australia, gipsy moth, Dutch elm disease and chestnut blight. We cannot get rid of these pests. I suspect that whilst many novel genetic organisms will be safe, some of them will not. If even a small fraction of them do become a problem we will have created biological pollution. These organisms may be alive and they could perhaps reproduce, mutate and migrate. We cannot easily recall them or eradicate them from the environment.

There is a science in chemistry called toxicology, a body of scientific knowledge used in risk-assessment analysis. However, in my opinion, even a super-computer will not be able to predict the ecological outcomes of the release of GMOs. Our ecosystems are too complicated. Therefore, the idea of working on models which are able to project the ecological impacts of any microorganism, plant or animal which is introduced into an ecosystem is over-optimistic, which is why the insurance industry will not countenance insuring against the risks of GMOs.

For those of you who are thinking that we are a risk-taking animal and that we can never foresee the outcomes even with scientific knowledge, I would say that the rule should be this: always intervene in nature on the side of prudence, intelligence and conservatism; never intervene radically and dramatically. It is because we cannot foresee all the consequences of our actions that we should always intervene intelligently and conservatively.

Is the deliberate release of thousands of genetically engineered novel microorganisms, plants and animals conservative, intelligent and cautious? Or is it radical, adventurous and dramatic? I believe we should not deliberately release any GMOs – be they microorganisms, plants or animals – into the ecosystem because the risks are immeasurable, and science – the 'technological fixer' – is no insurance against risk.

BOVINE GROWTH HORMONE

Bovine growth hormone (BGH) or bovine somatotrophin (BST) was introduced as a 'flagship' product for genetic engineering in agriculture and animal husbandry. It was to herald the emergence of a new era of genetic engineering. Four companies – Monsanto, Eli Lilly, Upjohn and Cyanamid – have invested at least US$1 billion in researching and developing this product. Imagine developing a genetically engineered product to provide more milk for the industrial world! Every industrial country is awash with milk and these four companies come up with the idea: 'Let's produce more milk.' This is a classic example of a limited definition of progress – more of the same. This product is bad for the cow, bad for the farmer, bad for the taxpayer, and probably bad for human health. It is only good for the profits of these four companies. Given that we are awash with milk, if this product is commercialised, up to one-third of all dairy farmers in the USA and in Europe will probably be out of business within 36 months. That would mean the devastation of entire rural communities.

As for the dairy cow, campaigners predicted at the outset that if you put growth hormone into an animal in larger measure than the animal is physiologically designed to metabolise, the outcome will be stress-related disease. Monsanto and the other companies said that this prediction was nonsense. Well, despite the fact that these companies fought hard to maintain confidentiality over the results of their BST studies, there are now reliable reports providing evidence of adverse effects – mastitis, stillbirths and other health problems – with these experimental cows. It is for this reason that this product has been held up by the regulatory authorities in the USA and the EC for years.

As early as 1986, consumer groups claimed they would prevent the commercialisation of BST products. I went around the world saying, 'This product will be "dead" on arrival.' Despite the enormous amount invested in its development, activists in Europe, the USA and elsewhere have kept it out of the EC, kept it out of the USA, and kept it out of Japan. BST is further away from the market-place now than it was 6 years ago. [In 1994 a moratorium was imposed on its commercialisation in the EU until the year 2000, while in the USA recombinant BST went on sale.]

THE NEW PARTICIPATORY DEMOCRACY

Most decisions that affect our lives are not made democratically. The computers, the televisions and products of the petrochemical industry affect our lives, our values, our politics and our identity much more than most of the decisions made in Parliament.

People make history. There is no *fait accompli*, no technological imperative which says a new technology has to be accepted. There is a younger generation – concerned and engaged – interested in empowering themselves. This is a healthy development for the human race. I believe that academic and corporate establishments are going to have to realise that the days of free, unencumbered, hierarchical decision-making are over, and that from now on there will be a multiple of constituencies that will effect a decision.

We are entering a new chapter in our relationship with technology: the public is going to become informed about, engaged in and responsible for decisions relating to science and technology before new products and processes come on-line. Participatory democracy will be extended into the last bastions of privilege, that is into science and technology. Genetic engineering will be the first technology in history to be publicly debated before it changes our lives, pro-actively, not after the damage is done. We have learned our lesson from nuclear and chemical accidents. Ask a lot of questions 'up front' and maybe our children will not be faced with the problems we are today.

PATENTING

In the USA, for the purposes of patenting, living beings are no longer distinguished from inanimate objects – a genetically engineered pig is no different from an automobile or computer.

Let me ask you a question, 'What is wrong with patenting animals?' If you do not intuitively know the answer to this question, I could probably never explain it to you. You may, of course, argue that we have kept animals as property since the neolithic revolution. Fortunately, however, the animal rights movement around the world is beginning to help us understand the need to move beyond such limited thinking. Moreover, there is a tremendous difference between ownership of an animal and the patenting of life forms.

Just as ownership of oil and mineral resources dictated who was to control the socio-economic system of the industrial age, whoever controls the gene pool controls the age of biology. Since UNCED a tremendous struggle has been taking place between 'North' and 'South'. Around 95 per cent of the world's germplasm is in the southern hemisphere, but economic and political power and advanced technology resides in the northern hemisphere. The MNCs of the northern hemisphere want to keep access to the gene pool open so that they can go down to Bolivia, for example, collect germplasm, bring it back, genetically manipulate it and sell it on the world market. Bolivia wants to enclose the gene pool so that it can get a royalty from its exploitation by MNCs.

However, I would like to say to both parties that the gene pool must be maintained as our common heritage. It is the biological 'parent' of all creatures on this planet. It is our biological and evolutionary destiny and should not be enclosed, commercialised or reduced to private property.

Here is a simple question to parents. Forget your institutional 'hats' for a minute. Will your children be better off growing up thinking of all living creatures as patented inventions? Of course they will not. Imagine the deprivation, the psychological alienation, the 'de-sacrosanction' of this form of reductionism. Are we willing to reduce the whole living planet to commercial property and, in the end, lose the spirit – the soul – of life?

RESPONSES TO ENVIRONMENTAL PROBLEMS

In the 1990s, as we start to make the transition into the age of biology – an age of environmental releases of GMOs, the introduction of BST, the patenting of genetically engineered creatures – as we deal with the tremendous destabilisation brought by five seconds of enclosure of the industrial age, there are going to be some alternatives for us to choose between. As the environmental and human crises deepen there are likely to be four responses.

1. 'It is not happening' – avoidance behaviour.

2. 'It is happening but I cannot do anything about it, this global crisis is so overwhelming.' Out of this response, this vacuum of cynicism, despair and resignation, we are going to see some very macabre political and religious movements emerge between now and the beginning of the third millennium. It is already happening in Europe and it is coming to the USA.

3. 'The quick fix' – somehow, somewhere, someone in Monsanto or Du Pont is going to find a way to suck the carbon out of the atmosphere and plug the ozone hole. Global warming, ozone depletion, species extinction, deforestation, desertification – there is only one way to fix it – genetically engineered forms of life. We are going

to come up with a plant that is drought-resistant, which can exist in, for example, desertified conditions in Africa; we are going to come up with an animal with a gene spliced into it that will be immune to ultraviolet radiation. If all that does not work, we are going to take these genes, tissue culture them indoors in vats, and create an indoor, hermetically sealed, enclosed agriculture so that we can avoid the problems of environmental pollution.

So instead of cleaning up the biosphere in order that it be conducive to evolution, we are going to leave all the pollution, all the 'side effects' of our technology out there and genetically tinker with the architectural design of life so we can grow our plants and raise our animals and our children to withstand the environment we have created. That, my friends, is pathology.

4. 'The leap of consciousness.' A leap of consciousness by one generation of parents and children who began to think, not as an American or an English person, but as a species. We cannot begin to deal with these crises and to structure the next technological age unless we begin to think as a species and to see the planet as a single biochemical entity. James Lovelock's Gaia biohypothesis [see Lovelock 1988] is just a rediscovery of what every traditional tribal society tells you in their ancient wisdom: we inhabit a single, living community; our biological rhythms are trained to the frequencies of this Earth.

How do we leap from national boundaries, technical boundaries and religious boundaries into 'species awareness' in one generation? A lot of people get lost at that point and say, 'Oh, come on, it takes time.' If I had come here 3 years ago and said: you will be able to buy the Berlin wall in department stores for US$9 a block; there will be one Germany; and the Soviet Union will just disappear from the European map, you would not have believed me. But we are not only seeing the end of the Cold War, we are seeing the end of a whole period of history. We are seeing 500 years of history being questioned on an unparaleled scale.

EFFICIENCY

There *is* another vision of the age of biology but to get to that we have to understand the roots of our present crisis. Why would people genetically engineer animals? Why would we tinker with the blueprint of life? Why would we use the biology of plants as a commodity? To understand this we have to understand the set of values, the principles and relationships that gave rise to the crisis we are in.

What key value has emerged from the last 100 years, a value that never existed in its modern guise 100 years ago, a value which is the key to our science, underwrites our technology, is critical to our economical policy, and motivates our private and public lives? How about this as an answer – 'efficiency'? What do you think genetic engineering is about? It is all about efficiency. Maximise your output in the minimum time. Reduce the minimum time and increase value in the process.

One way of looking at genetic engineering is the application of the standards of market efficiency to the natural world. Get that pig leaner, make it grow faster, get it to market quicker. Make sure that tomato does not decay on the shelf for another 6 months so you can have efficient marketing management to sell to the consumer. Do you see what I mean?

Would you be surprised if I told you that efficiency is a new value? Look in Samuel Johnson's dictionary: you will find that 'efficiency' is defined as having to do with effecting divine causes. It only became a market value term as we know it with the introduction of thermodynamics, mass-production and the division of labour.

Could an 'efficient' environment ever be a sustainable one? How many of you have been to Sienna in Italy? A beautiful city, as I am sure you will agree. Go up that clock-tower and look over the whole architectural blueprint. The city is 1000 years old and it took a tremendous amount of labour, energy and capital to build it. It has lasted 1000 years. But how many American or British suburbs will be here 1000 years from now?

Is there an alternative to efficiency in animal husbandry? Remember, gene splicing is designed to make the *most* efficient genetically engineered animal. Is there an alternative to efficiency? And if so, what would it be?

Can an efficient environment ever be playful, joyful? Has anybody here ever worked in McDonalds? Has anybody ever worked in a hyper-efficient environment that allowed you to blossom? No. What about efficiency in personal relationships? 'Honey I love you, so I'm going to maximise my affection for you by being efficient in our love-making.' An absurd idea. So, why do we use this value to orchestrate every one of our scientific, technological and commercial relationships?

WORLD VIEWS

Between the 1620s and the late nineteenth century European scholars created a new 'world view', a new way of structuring our thinking, our cognition and our relationships. To understand this world view is to understand the choice between a genetically engineered future and an ecological one. Genetic engineering is the final extension of Enlightenment thinking – the capitalisation of the biology of the planet. It is not a new vision. It is the last desperate gasp of the old vision of progress.

Francis Bacon took on the Socratic scholars, the Greeks, the medieval schoolmen and the Church, and said 'Look, you're too interested in why, I'm interested in how.' He said we could detach ourselves from nature, become a neutral observer, force nature to do what we want it to do. Knowledge is power. Francis Bacon called nature 'a common little harlot, whose wildness needs to be moulded, squeezed and subdued.' That is the objective of modern science. Once the concept of the neutral observer is introduced, we detach ourselves from any relationship with the other. It becomes a subject/object world. Control, power and capital in business.

Let us consider personal relationships. If I attempt to squeeze, mould and subdue the person I am with to make them conform to the way I would like them to be, does that relationship grow? Why, of course not. The scientific method is all about the structuring of relationships in the world. And we wonder why we have a world in crisis, a world which is polluted.

Does this have anything to do with genetically engineered pigs? John Locke, the eighteenth-century English philosopher, was the ultimate political philosopher. John Locke said that everything in nature is waste because it is unenclosed, it is not reduced to property, it is not doing anything for you. He said it becomes valuable when you turn it into products.

Let me suggest that many people no longer believe this world view. Everything in nature is not waste; it is valuable. Intrinsic value comes first, utilitarian and thermodynamic value last. We simply take nature's intrinsic value and transform it into goods and utilities, products and services. And what happens to all of them eventually? They end up as waste – entropy.

If you believe in world views then what is gross national product (GNP)? Our economists say it is a measure of national output, the goods and services produced every 12 months. What do you think it is? It is a measure which places a value on the temporary products we produce by transforming and exploiting the Earth, a process which creates waste, and ultimately the goods themselves become part of the waste stream. It is a measure of exploitation. And yet we have the whole world still talking about increasing GNP!

What am I trying to get at? Everything is borrowed. My tie is not going to be a tie in 100 years from now. Where did it go? This building is not going to be a building 1000 years from now. Where did it go? The fabric and fibre of our being, the accoutrements we create, the technology we use, the monuments to our immortality, are all borrowed. Everything starts off with nature, which we transform for our temporary use and exploitation, and it goes back whence it came. The *Bible* is clear on this subject. 'From dust to dust, ashes to ashes.'

We call this the 'age of growth'. The American presidential candidate, H. Ross Perot calls it the 'age of borrow' because when you borrow you have an obligation to pay back. Debt is a relationship of mutuality, of give and take. What if I could wave a magic wand so that every textbook in the world would eliminate the word 'growth' and substitute the word 'borrow'. Would that change the thinking and world view of a generation? Would we have global, environmental and human crises? Would we be engineering biological blueprints for short-term market efficiency? Would we patent life?

Owen Barker identified three stages of consciousness. The first was a deep, intimate participation with nature with no sense of self and no violation. The second stage of consciousness was from the neolithic revolution onwards during which we detach ourselves from nature and control it at a distance. We gain a sense of self and volition but we lose intimacy with the body power of nature. Barker suggests that we may be on the brink of the third stage of consciousness, a stage in which we again participate with the spirit and biology of the planet, not out of fear as primitive hunter-gatherers, but a stage of self-aware choice.

In this new philosophy, science, technology, progress, economics and politics may provide an alternative vision for the activists and scientists. It may provide an option very different from genetic engineering. The first signs of this third stage are the development of a new philosophy of science based on empathy, relationship and context rather than detachment and neutral observer status.

Can one be in favour of science and still be respectfully critical of the narrow methodology we have been using since the Baconian revolution? Am I saying, 'Throw out the scientific method of Bacon?' No. What I am saying is that it is alright to have Francis Bacon's methods [see Bacon 1624] if they are balanced by a new, relational, alternative methodology for exploring relationships. Let us see if we can design another method for science that is rigorous and ethical, that gives us the possibility of knowing why and not just how, so that we can readjust ourselves to the scheme of things. Was the Earth made for us or were we made for the Earth? We need a new philosophy of science based on empathy and relationships.

How about a new concept of technology? There are two great 'cop-outs' used this century. One is, 'I was only following orders'; or 'We can't do anything about it.' The second is, 'If it can be done, it will be done.' Tools are not neutral; there has never been a neutral tool in history. The idea that tools are neutral and only have to be regulated is flawed. The only question that is ever asked about genetic engineering is, 'How shall we regulate it?' How about, 'Do we need it?' Or what about, 'How will it change our concepts about relationships?' With hindsight, did we need the automobile, a technology with a devastating effect on the biosphere which has killed more people than all the wars this century? So that we can travel to work in central London at the same speed as we did 50 years ago? Half the trips we take in the USA, for example, are only 3 miles long.

Technology is not neutral because it is power. All tools are power and power is never neutral. A bow and arrow gives you more power than your throwing arm, a locomotive or jet or car gives you more power than your running leg. Computers amplify memory. Tools are power. They extend our biology, our natural limits, so we can overcome space and compress time in order to expropriate. That is what a tool is.

The question before the next generation with regard to biology is, 'How much power is appropriate for intervening with biology?' 'Is the relationship between the power of the new technologies and our relationships with the scheme of things and the needs of future generations inordinate or inappropriate?'

We need to adopt technologies that are sustaining rather than draining. Does genetic engineering drain agriculture and animal husbandry? Does it suck up and consume in the service of the market or does it sustain the ecosystem for future generations? Is it driven by expediency or by sustainability?

If it is possible to define a new science based on empathy and relationships, and a new technology based on appropriate power and sustainable technology, how about a new concept of progress? We defined progress, when I was growing up, as more output in less time, a definition which leads to silly products like BST. Perhaps a more sophisticated definition of progress would be initiatives which enhance the well-being of the community, protect resources, and provide for the needs of future generations of humans and other creatures. Parents, which definition of progress would you prefer for the future of your children and your children's children? Would genetic engineering comply with the new definition of progress? No. Not only do we need a new concept of science, embodying empathy and relationships; a new concept of technology embodying sustainability and appropriate exercise of power; but also a new concept of progress, which takes the needs of present and future generations into consideration.

The environmental movement is split. There are a lot of environmental activists who think that everything is a resource. What is wrong with believing that all life is a resource? It puts people in the same camp as those who think that everything is a utility that can be reduced to commercial property. Are all plants and animals here just as a resource? Are they only here for our needs or do they have some intrinsic value, some right to the full blossoming of their being, independent of our needs? I am not saying we should never use and need other creatures, but there is a difference between respectful give-and-take, and instrumental exploitation which perceives every living creature solely as a utility for profit, consumption and short-term market expediency.

Does genetic engineering respect the intrinsic nature of other creatures or does it reduce them to 'chemical factories'? Will your children be better off or worse off growing up in a world thinking of animals as chemical factories that bequeath pharmaceuticals?

If we need a new concept of science, a new concept of technology, and a new concept of progress, how about a new concept of economics? There have been two basic kinds of economics in human civilisation: the first is the ancient economics of 'generativity'; The second is the modern economics of 'productivity'. The two are very different. Ancient economics was the economics of gift-giving; modern economics is the economics of caveat and tort. 'Generativity' is a life-affirming principle; 'productivity' is a death principle.

If you wonder why we are moving towards a death culture it is because our interactions with the biology of the planet are based on productivity. Production is a matter of exhuming, transforming, discarding and exhausting. Consumption, if you go back to the Old English Dictionary, means to lay waste, to destroy.

We only practise ancient economics one day a year now – at Christmas – and then with such an orgy of guilt and a sense of incomplete relationships that we have suicide, depression and family arguments. We cannot repay everything in just one day a year.

If we need a new science of empathy; a new technology of appropriate powers and sustainability; a new concept of progress as providing for the rights and needs of the community of future generations; and a new concept of economics as generative rather than productive; how about the new concept of ethics?

What are the ethics of producing genetically engineered animals? Our problem in addressing this question is that we are used to dealing with 'common' evil based on Christian theology: you murder someone, you go to jail. But in the modern age we have 'cold' evil as well as 'common' evil. If you look at Dante's Inferno, the twelfth circle of hell is ice. Rudolph Steiner, the German mystic and Christian philosopher, talks about 'cold' evil, which is evil so mediated by technological and institutional barriers that we, as beneficiaries, barely perceive that we participate in the victimisation of another person, being or thing. A child eats a cheeseburger at McDonald's: 'cold' evil. How could that child experience the same feeling of having perpetrated a violation as if he had stolen something? What is the price of a hamburger? We have got 1.2 billion cows on the planet. They take up 25 per cent of our land surface, they eat nine times our calorific content, they weigh more than the human race, and if an anthropologist came down from another planet he would think they were the dominant species. They are one of the top ten environmental threats on earth. They are a major reason we have deforestation in Mexico and Central and South America; they are a major factor in desertification; they are destroying millions of acres of public land in the USA, and West and Central Australia. Cattle are a major source of groundwater pollution in the Netherlands and the Mid-west of the USA. They contribute to water scarcity because of the tremendous amount of water used to irrigate their feed crops. They even contribute to global warming – not so much through the methane they produce, but the carbon dioxide produced when the Amazonian rain forests, for example, get burned, in part to provide land for pasture.

Beef production is a major problem for human health. We are dying of cancer, we are dying of strokes and we are dying of heart attacks and other diseases of affluence. We are dying of diseases of affluence because we are gorging grain-fed beef. Here is the irony – a billion people will go to bed hungry tonight whilst one-third of all the grain produced in the world today is going to livestock in Europe and Japan. Thirty years ago the peasants were on the common land in Mexico and Central and South America. They were growing corn and beans. These commons have been enclosed and the peasants have been removed. They have lost their right to the land. Now they are growing soya and shipping it to Europe for livestock feed. We die of diseases of affluence from grain-fed beef whilst they do not have any grain. The hamburger is 'cold' evil: our health suffers, the planet's environment suffers, the poor do not have enough arable land to grow their own food, and there is also the issue of the inhumane treatment of the animals themselves.

In the 1990s we are going to see a tremendous transition in the animal rights movement. Their concern for laboratory animals will continue but will broaden to include the welfare of domestic farm animals – the millions of pigs, sheep and cows – which are amongst the most abused, the most maligned and the least regulated species on the planet. Showered with insecticide, pumped full of hormones, cramped into trucks, having their backs and hips broken, and then being slaughtered in primitive conditions – is this what we want in the twenty-first century? We want to colonise outer-space, yet we still have not civilised our own 'garden'. How do we tell a teenager that eating that cheeseburger is 'cold' evil?

Do we reduce all life to technology? Do we genetically engineer all farm animals? Do we turn them into chemical 'factories' for the production of pharmaceuticals? Or, do we restore our proper relationships and recentralise our values with other creatures? I predict that by the twenty-second century there will no longer be domestic farm animals on this planet. We are in the twilight years of beef culture. We will not be able to afford an environment where a handful live high up on a protein chain at the expense of poor people, other creatures and the environment. The question is whether we understand this and can respond to it in an enlightened way or whether we resist acting now and attempt to deal with the problem only when catastrophe occurs.

PURE FOOD BEYOND BEEF

I believe that the politics of food will be the entry point for the new politics of the decade. The kind of deep ecological vision that has been brewing within the feminist group, the animal rights group, the environmental movement, the sustainable agriculture movement and the health movement is going to politicise food. Food politics are going to dictate our approach and the doors will open to the age of biology.

In May 1992 in the USA, Vice-President Dan Quayle and the Food and Drug Administration (FDA) announced the

USA government's policies on genetically engineered foods. There will be no premarket testing, no labelling, and, with few exceptions, the companies will police themselves.

In opposition to this, we have announced the 'Pure Food Campaign'. We are determined that genetically engineered foods will not reach the market. Remember, in 1986 I told a lot of people from the transnational companies (TNCs) that we will ensure that BST is 'dead on arrival'. Now I would like to say something else. Genetically engineered food is going to be the greatest financial disaster in food technology history. You are going to see a mass of consumer resistance throughout the EC, the USA, Japan and other countries around the world. You see, genetically engineered food is part of the old vision of the World Fairs of the 1930s and 1940s. Today we have a new generation that wants organic foods, sustainable agriculture, ecological living. They do not want chemicals, pesticides and hormones, and they do not want genes in their corn, firefly genes in their rice, and human genes in their asparagus. You know why? Not because they are slightly unprogressive or conservative but because there is a new sense of responsibility towards the environment and to animals. We are going to see a very strong coalition. We have already begun to put together a Eurogroup campaign of farmers, environmental activists, animal rights activists, consumer health activists and Third World development campaigners.

'The Pure Food Campaign' is pressuring for the premarket testing and labelling of all genetically engineered foods in every country. We have announced a permanent, irreversible boycott of genetically engineered food because we want a new science of nutrition, we want permaculture and we want organic farming.

In the first 2 months of this campaign, 1500 of the greatest chefs in the USA – that is, virtually every great chef that I know of in the USA, in every great restaurant – have joined the boycott. They are putting the new international seal of the 'Pure Food Campaign' on their menus. The seal is a double-helix with a slash through it (see Figure 2.1). We do not have genetically engineered foods on the menus of the greatest restaurants in the USA.

Now we are going after the growers, the wholesalers and the retailers – the subsidiaries of the food industry – to bring

Figure 2.1 Seal of the 'Pure Food Campaign'

them with us into the 'Pure Food Campaign'. By the mid-1990s, the 'Pure Food Campaign' will be all over Europe.

The second campaign we are launching, in April 1993, is the 'Beyond Beef Campaign'. We need to lower the position of the eating habits of western civilisation on the food chain, and we are asking people to reduce their beef intake by at least 50 per cent and replace it with grains, fruits and vegetables that are organically raised. We are targeting the fast food industry – businesses like McDonalds.

I believe people are basically receptive when they get the information and when they get it in an engaging rather than confrontational fashion. I believe that the campaigns against genetically engineered food and over-consumption of beef are going to be unprecedented successes.

I am going to leave you with some words of the Iroquois Indians. When the Iroquois had to make decisions they first called on the elderly women because they were considered to be the carriers of wisdom – how times have changed! If the elderly women could not resolve the dispute the tribal chiefs would get around a fire and ask this question. 'How does the decision we ponder on today affect seven generations' renewal? Our children's children's children's children's children's children's children'. They heard their ancestors speak from the grave; they also heard the unborn children and all the other creatures not yet born saying, 'Look out for our interests, we are not here yet'. So when the Iroquois made a decision they did so in part with great deliberation of history and of the future time and space. If we thought like the Iroquois would we be tinkering with, rearranging, manipulating the genetic code of farm animals and plants and humans for short-term gain? Many young people look at my generation and say, 'Boy, you really screwed it up and then you preach at us to fix it.' I hope that when the children of this generation look back, they will say, 'My mother and father had the courage to criticise the old vision.'

I have great respect, by the way, for scholars who attempted to create a new world. We should not throw out all that sort of thinking but I think we should gingerly and respectfully temper it, modify it and use only those aspects of it that are still usable. I believe there must be a continuity with history, not a radical break. It is not enough to have new ideas, we have to have action; but it is not enough to have action devoid of rigorous intellectual scholarship because then it could be even more irrational than previously. So we need a generation to say, 'We will look at the old vision. We will create a new way. We will grow toward it. We will commit ourselves to self-empowerment. We will rebuild the Earth. We will re-establish our relationships. We will create more humane standards. And we will live in harmony with the other creatures.'

3 Transgenic Farm Animals and Enhanced Productivity

Dr Kevin Ward

The ecological crisis that is now facing the planet is the result of a large and rapidly increasing human population showing a disregard for the basic principles of sustainable ecology. While the population remained small, the wasteful use of natural resources and the introduction into the environment of toxic pollutants could take place with little observable impact on the biosphere. With a population now exceeding 5 billion individuals and increasing exponentially, these same practices cannot be further tolerated. Already there are appearing significant signs of environmental stress which, if left unheeded, will precipitate a global crisis that threatens the very fabric of modern civilisation. Such a situation is avoidable, but only if we take immediate steps to correct current abuse and put in place practices that will maintain renewable resources at current levels and minimise the depletion of non-renewable resources.

Just as the causes of the current environmental crisis are complex, so too are the solutions. It is mandatory that some form of control be placed over the exponential growth in the population because we are rapidly approaching the finite limit to the carrying capacity of the planet. A reduction in the use of non-renewable resources is necessary and this must also be coupled to a programme of resource replacement. There must be a halt to the constant introduction to the environment of pollutants that degrade its ability to support life. Moreover, it will also be mandatory to make significant changes to current agricultural practice.

Agricultural efficiency has steadily improved in the 10 000 years since its estimated introduction to the support of human society. Much of this increase has been the result of improved farming practice which includes better ways to manage crops and animals, protect them from predation and disease and collect the resultant products. These improvements to agricultural efficiency through the route of improvements in farming practice have not been without cost. It takes much more energy input to produce one calorie of output from modern high-yield agriculture than from traditional agriculture. For example, to produce a calorie of food as beef uses 0.5 calories of input energy when produced by traditional range-land feeding and 10 calories of input energy when produced by intensive feed-lot methods, a 20-fold difference. This only serves to highlight the problem we must eventually face that although the production of food has so far maintained itself in comparison with the growth in population, continued increases from improved management alone are unlikely to be sustainable in a world about to enter a period when energy is less freely available.

The second route to improved agricultural productivity is to enhance the genetic potential of cultivated plants or animals. This is achieved by selective breeding of superior individuals and has proven highly successful as evidenced by the superior performance of today's cultivars and breeds compared with those of past eras.

For domestic animals, productivity in selected areas such as milk yield, wool production or carcase quality can be steadily improved by selective breeding, rates of gain often being 2–3 per cent per annum for many decades. However, this approach suffers from two significant disadvantages. Firstly, as yields constantly improve, it can become increasingly difficult to identify animals with production characteristics significantly different from the mean of the population. Secondly, and perhaps more importantly because the unit of genetic exchange is the whole chromosome which contains many thousands of genes, it is often difficult to separate a desired productivity characteristic from one or more undesirable characteristics. Associated with this is the inability to transfer characteristics between species because it is not possible for different species to interbreed successfully.

The recent development of genetic engineering has provided a technical solution to this problem. This technique allows for the transfer to any domestic animal species of small pieces of DNA which have been characterised in very fine detail, usually all the way to the complete base sequence which is the ultimate definition of a gene's structure. This technology has a number of significant advantages over conventional breeding. First, it provides a way for genes of good effect to be transferred without the simultaneous transfer of deleterious genes. Secondly, and probably more importantly in the longer term, the technique provides a mechanism for the transfer of genes across wide distances of biological separation, so that essentially the entire resource of genetic information in nature is available for redistribution between species as appropriate. Used with judicious care, this provides a mechanism for the long-term improvement of agricultural productivity in a way designed to conform to the restraints that are inherent in a new era of sustainable agriculture in an energy efficient, non-polluting society.

GENETIC ENGINEERING THROUGH EMBRYO PRONUCLEAR INJECTION

It has been known for about 20 years that isolated DNA could be introduced into cultured mammalian cells by a variety of methods and that in some instances this DNA became a stable part of the cell's genetic information. The level of integration in these experiments was very low, of the order of one integration per 1000 cells, and this excluded the technique from being used to manipulate animal embryos. However, it was subsequently discovered that if the foreign DNA could be injected directly into the nucleus of cells the frequency of integration increased dramatically. This immediately provided the stimulus for a number of groups to consider the transfer of DNA into single-cell animal embryos by injecting the DNA directly into the nucleus using techniques called embryo pronuclear microinjection.

In 1980 and 1981, several groups virtually simultaneously announced that this approach was indeed successful in providing a method for the stable integration of foreign DNA into mouse embryos, giving rise to 'transgenic' animals. Simply described, the technique consists of obtaining from a donor animal some single-cell fertilised embryos, placing them under a microscope fitted with special optics and inserting into the nucleus of the embryo a fine glass needle containing the DNA to be transferred. The microinjection process is usually controlled by a special pneumatic pump which can deliver the minute quantities of DNA required (about 1-2 picolitres, or about 1 millionth of the amount that can be fitted on the head of a pin). After a period of recovery, the embryos are then placed into the reproductive tracts of recipient females where they continue their development to live animals. Not all animals born from these embryos retain the new gene, but the process is good enough to enable most modern laboratories to carry out the technique once they are suitably equipped.

The application of this technology, which was developed in mice, to domestic animals has not been simple because of a number of technical problems associated with the embryos themselves. Apart from the inherent difficulties associated with the production, collection and transfer of embryos from large animals compared with laboratory mice, it was quickly discovered that most domestic animal embryos contain in their cytoplasm a large number of dense granules that tend to obscure the nucleus and thus make it difficult to insert the microinjection needle into the desired position. This problem has been overcome in different ways depending on the species involved, and the technique of embryo pronuclear microinjection has now been demonstrated in all of the major domestic animal species, albeit with success rates considerably lower than those obtained when using laboratory mice.

EMBRYO PRONUCLEAR MICROINJECTION FOR ENHANCED AGRICULTURAL PRODUCTIVITY

There are a variety of potential applications for this technology to improve domestic animal productivity. However, it is an indication of the difficulties associated with the technology that to date there is no example of a transgenic animal so produced which is successfully deployed at a commercial level in the agricultural industry, although several projects are close to entering this phase. The areas of potential application include:

1. The modification of animal biochemistry;

2. The manipulation of the endocrine system;

3. Improving disease resistance;

4. Alterations to important structural proteins;

5. Manipulation of secretory gland function, particularly the mammary gland.

Some examples of the current research in these areas are summarised below.

THE MODIFICATION OF ANIMAL BIOCHEMISTRY

One of the less desirable properties of the evolutionary process is its inability to conserve for future use genetic information that confers no immediate selective advantage to a species. This has resulted in the loss of a number of biochemical pathways in our common domestic animals as evidenced by the range of essential amino acids, vitamins, lipids and cellular co-factors that must now be supplied by diet instead of being synthesised by the animal. Whilst it might be argued that random mutation and natural selection could regenerate missing information that was of sufficient importance to the survival of a species, this is in fact most unlikely. The genes that encode the fundamental biochemical mechanisms currently used by all organisms on the earth were probably produced about 3 billion years ago in a relatively short period of geological time, a period which may have been as short as 100 million years. This was possible because of conditions that existed at that particular time on the planet which greatly favoured the growth of single-celled organisms with their short generation times and high rate of mutation. Since then, these genes have been modified in a variety of ways to give us the complex array of life we see today, but once a particular gene has been lost from a species, it is essentially lost forever.

One of the most powerful uses for genetic engineering of domestic animals would be to increase the supply of nutrients currently limited but which would be of advantage. The general approach is to isolate functional genes from a bacterial source, modify the genes for expression in the target species and transfer them by standard procedures of embryo microinjection. In Australia the attempt is now under way to introduce the cysteine biosynthetic pathway to sheep in order to improve wool growth in periods of reduced feed supply.

The biochemical pathway involved is:

acetyl-CoA H_2S
serine ———> O-acetylserine ———> cysteine
serine transacetylase O-acetylserine sulfhydrylase

The relevant genes (*cysE* and *cysK*) have been isolated from the bacterium *Escherichia coli*, fused to a eukaryotic promoter and transferred to mice and sheep. In transgenic mice, the new biochemical pathway has been shown to be fully operational in several key tissues, including the intestinal tract and the liver. These transgenic mice remain perfectly healthy even when the pathway has been operational for most of their lifetime, clearly demonstrating that in this case at least it is possible to alter the species' biochemistry without obviously disturbing the animal's homeostasis. Key experiments with transgenic sheep are planned with the full expectation that in this species too the pathway will be functional but not harmful.

THE MANIPULATION OF DISEASE RESISTANCE

The manipulation of the immune system in transgenic animals is discussed elsewhere in this book (see Chapter 4). However, a separate and unique approach to the modification of disease resistance is the research directed at allowing animals to produce specific proteins which might provide protection against a variety of insect pathogens. This project, still in its early research phase, is investigating the possibility of using the enzyme chitinase (see later) to inhibit a range of insect pathogens. Specifically aimed at the inhibition of the sheep blowfly, this approach also has the potential to attack other insects which cause livestock problems such as lice infestation in sheep and tick attack in cattle.

The principle underlying this approach is that the insect cuticle (outer covering) and a portion of the insect digestive tract contain chitin, a biopolymer made of *N*-acetylglucosamine units. The enzyme chitinase specifically degrades the chitin biopolymer with deleterious consequences to the insect. We have made a genetic construct which comprises a gene encoding chitinase together with the promoter elements necessary for the expression of this gene in the skin of sheep and cattle.

Our hypothesis is that the integration and expression of this gene in an animal should provide a measure of protection against insects that ingest the chitinase or whose cuticle is sufficiently sensitive to the enzyme to be degraded over a period of some hours. Our initial results support this view but the project remains in its early phase of research and we still have a substantial amount of preliminary work to do before considering the transfer of the gene to domestic animals.

The environmental attractiveness of this project stems from the high specificity of the enzyme for insects, its total lack of toxicity to all mammals and, in the case of attack on blowfly larvae, its limited impact on the total population of the target insects. Currently the problem of blowfly strike in sheep is controlled by the use of insecticidal chemicals which can cause significant environmental problems in terms of residues left on the wool and in pasture contamination. The genetic engineering approach provides an attractive alternative with the possible added advantage that it will be difficult for insects to develop resistance to the enzyme since this would involve making a fundamental alteration to the structure of the cuticle.

MODIFICATION OF ANIMAL PRODUCTS

One of the most interesting possibilities for the use of transgenic manipulation is to produce transgenic animals with genes capable of modifying important animal products such as wool and milk. The approach is simple in concept and involves the modification of the gene encoding one of the structural protein components of the target product in order to change product quality. For example, it may be possible to alter many of the important physical properties of wool by making relatively simple changes to a few of the key keratin proteins that make up the mature wool fibre. Many of the genes encoding these proteins have now been isolated and characterised, thus opening the way for experimental manipulation.

It may also be possible to manipulate the composition of milk to improve its nutritive value for human consumption, for example, by making the protein component more homologous to human milk by altering the lipid (fat) or carbohydrate profile. This has some similarity to the research to produce novel proteins in milk (see Chapters 7, 8 and 9), but its goal is the production of a food for consumption rather than a high-value pharmaceutical product.

ETHICAL AND SOCIAL IMPLICATIONS

This topic receives considerable attention in other chapters of this volume but is of such importance that it should also be considered in the context of the issues discussed here. The questions raised under this heading are extremely complex and involve fundamental moral issues. Of particular relevance is the question of public safeguards against abuse of the technology for short-term commercial gain or for deliberate social engineering, but it is also important to consider the deeper philosophical questions related to mankind's licence to alter the fundamental genetic composition of other species.

Public safeguards are of paramount importance in a technology with the power of genetic engineering. The extreme position is of course a complete ban on its use, regarded by some as the ultimate safeguard. I hope from the information provided in the earlier part of this paper that advocates of such action might reconsider their views in the realisation that not only are the potential benefits from judicious use of the technology of great value but also that they may hold part of the key to the long-term survival of human civilisation. At the other extreme is the situation where no control exists over the application of the technology. This is clearly untenable and is recognised to be so by essentially all scientists involved in such work. The correct position is perceived by most of the public as lying somewhere between these extremes and it is the identification of exactly where this position is that is currently the subject of intense national and international debate. The difficulty often revolves around the use of voluntary guidelines for the release of genetically engineered animals versus the release only under strict regulation governed by legislation. It would be in my opinion a fair summary to say that most scientists believe voluntary guidelines are sufficient to ensure safety while the public generally favours legislated regulation.

In Australia, a comprehensive enquiry established and conducted by the federal government to examine the issues inherent in the release of genetically manipulated organisms (GMOs) has recently been completed. A number of important recommendations have been produced from this, with the need for legislation to control the release of GMOs clearly identified. The necessary statutes are currently in preparation and are expected to be in place very soon. It is of interest to note that the reaction of the scientific community in Australia to this report has been very positive and the need for legislation accepted without dissent. Furthermore, the majority of the lay community also appears to be in agreement with its findings, although there are some who still retain significant reservations that the legislation will be sufficient. There is little doubt that Australia will be a useful test for the value of the technology because of its strong dependence on agriculture and its general acceptance of the concepts of genetic engineering.

The question of whether or not mankind has the right to manipulate animal genetic properties in any way at all involves the consideration of much deeper moral issues and is unlikely ever to be fully resolved. It cannot be denied that genetic manipulation by conventional methods of selective breeding has been practised by mankind for millennia and in general is accepted as a legitimate pursuit. The difficulty therefore seems to arise because the deliberate introduction to a species of a piece of DNA which could not by natural breeding have been transferred is seen by some as contravening nature's laws. This is compounded by the fact that scientists do not as yet possess a complete understanding of the complex interactions that regulate the activity of the many genes possessed by animals. However, this is a superficial objection that will eventually not be relevant as our knowledge of gene regulation advances.

On the other hand, the view held by those who believe it is acceptable to transfer genetic information between species is generally that the genetic resource of nature can be considered a single entity, albeit with a range of gene combinations which engender a constant admiration for the evolutionary process. While the biological diversity of the planet must be preserved as a priority, the concept of a single fundamental genetic resource allows for the movement of genes which would appear to be of advantage to a particular species but are not available to it by conventional breeding because of past evolutionary history.

It is unlikely that the two opposing views can ever be fully reconciled. Those who hold that the essence of a species lies in its genetic information and that any modification of this creates a new species will inevitably have difficulty accepting the widespread mobility of genes across nature. However, it is clear that mechanisms already exist for such transfer to occur naturally in bacteria, plants and insects. Furthermore, the presence of a range of repeated sequence elements in the genomes of vertebrates suggests that at some time during the evolution of animal species, widespread transfer of DNA between species was possible.

ADDRESSING PUBLIC CONCERNS

It is clear that the technology of genetic manipulation is now firmly established and provides a method for the manipulation of the genetic properties of domestic animals. The laboratories involved in the work believe it has much to offer in the enhancement of agricultural productivity and this view is shared by the majority of the agricultural industry. It is also clear that the technology is of such potential power that it creates public concerns at many levels. These can only be addressed by discussion between all parties in the hope that common ground can be found. Legislative control is inevitable and a number of countries are working towards the establishment of acceptable laws to govern the work. It would be desirable but logistically difficult to establish an international regulatory framework for the new technology. The Organisation for Economic Co-operation and Development is working towards such a goal but it would be naive to expect consensus between all nations to come easily. While such moves are in progress, the establishment of efficient communication channels between nations should be a priority. Let us hope that this volume will be an important step towards such communication and ultimate international cooperation.

4 Genetic Engineering for Disease Resistance

Dr Ivan Morrison

ANIMAL DISEASES

In considering diseases of farmed livestock it is important to note first that diseases are not merely 'nuisances' that hamper the efficiency of livestock production but in many instances they are a major cause of death and suffering in the animals in agricultural systems. Second, that this is true not only of our modern, intensive agricultural systems but also of the more traditional systems employed in organic farming or developing countries where the diseases may be different but, nevertheless, they impose similar problems for production and indeed for animal welfare. The third point is that for many of the major diseases of farm livestock the means we have of either preventing or treating the diseases are really quite inadequate at the present time, and so there is clearly a need, both from the viewpoint of improving productivity and for improving animal welfare, for alternative approaches. This chapter provides a few examples of how biotechnology can contribute in a positive sense to addressing the problems caused by livestock diseases.

The two areas discussed are: first, the use of DNA technology for the production of new generation vaccines; and second, the use of genes to create animals which exhibit a greater degree of genetic resistance to disease. While productivity is important in terms of efficient production of food of high quality, there are a number of other issues that are of equal importance in considering animal disease. First, animal welfare, which is inextricably linked to productivity. Many of the diseases which cause a constraint in productivity can also cause considerable suffering in the animal community, and in certain instances one could argue that by improving disease control one could perhaps produce similar quantities of food from a smaller breeding stock – a change of particular relevance in developing countries where the number of agricultural animals presents a threat to the environment.

Another important issue is consumer safety. Many of the potential new vaccines or genetically manipulated disease-resistant animals will probably lead to improved consumer safety in terms of the foods that are derived from these animals and, in certain circumstances, improved environmental safety. There are, again particularly in developing countries, diseases which are currently controlled extensively by the application of veterinary drugs, and clearly if there were alternative methods of disease control one could rely much less on the use of such drugs and hence avoid the potential environmental problems that they pose.

The third important issue is sustainability. The introduction of both vaccination and genetic resistance to disease offers tremendous opportunities to produce much better sustainable systems than the current practice using veterinary therapeutics.

An important point about biotechnology, whether one is talking about vaccination or genetic resistance to disease, is that it is generally based on detailed scientific understanding. In many instances the traditional approach to developing live, attenuated (weakened) pathogens for use as vaccines was based on empirical *in vitro* selection procedures which resulted in live vaccines that were, in fact, genetically manipulated organisms. These organisms were then distributed into the field without any consideration of the nature of the genetic change and how it affected the biology of the organism. By contrast, the use of DNA technology to modify pathogens genetically in order to develop live vaccines involves studies that lead to a detailed understanding of the pathogen in terms of how it functions in the host animal and of the particular genes of the pathogen that cause disease. This understanding allows one to alter the organisms in quite a rational manner. Moreover, research on a new generation of vaccines will rely on an increased understanding of the immune responses of the host, and therefore an improved general understanding of the relationship between the pathogenic organism and its host.

GENETICALLY ENGINEERED VACCINES

The types of vaccines that can be produced with recombinant DNA technology fall broadly into four categories. The first of these involves the identification of individual proteins in the pathogen, for example a virus or parasite or bacteria, which are the targets of the protective immune response of the animal. The next step is to clone the gene that encodes the protein of interest, express the protein and then use that protein as a vaccine. In this approach, one is merely applying a gene product to the target animal, rather than a gene.

A second approach is to generate attenuated (weakened) organisms through, as described above, the application of a basic understanding of how the organism functions in its host. The pathogens are attenuated by removing genes that are involved in the production of disease in the host. This represents a substantial improvement on previous, rather more empirical, approaches and can be quality-controlled in a much more precise manner.

The third approach involves the use of these attenuated organisms and is, perhaps, somewhat more emotive. The principle applied is that if one has an attenuated virus or bacterium, one can then insert genes from other organisms into the sites of the genes that have been removed. Such manipulation applies an understanding of the function of these genes, and clearly requires strenuous quality testing and stringent safeguards against the possible hazardous effects of these organisms. However, such organisms have the potential to make a major impact in controlling infectious disease.

The fourth approach relies on the potential of recently developed technology to construct the outer shell of a simple virus. We have the means of generating a vaccine which, although it has been produced *in vitro* by recombinant DNA technology, does not contain any genetic material. In other words, the genetically engineered vaccine would be the outer protein shell of the virus, without the genetic information inside it.

A number of examples of how these various technologies have potential application follows. First, the use of proteins as vaccines. In many tropical and subtropical regions of the world, using traditional extensive livestock systems, ticks pose a major problem. They suck blood from the animals causing them substantial distress, and they also transmit a number of important infectious diseases. In Australia, the strategy which has been pursued to try to immunise livestock against ticks has been to isolate molecules from the gut of the tick and use them to produce anti-tick antibodies. The livestock animal then imbibes blood containing the anti-tick antibodies which results in the destruction of ticks that subsequently infest it.

The next example is of the second and third approaches to vaccine production, namely the use of attenuated pathogenic organisms and genetically modified forms of these organisms. The example involves the capripox viruses which infect goats, sheep and cattle in the tropical and sub-tropical regions of Asia and Africa. There is widespread distribution of these viruses over the African and Asian continents where there are a large number of other infectious diseases for which there are inadequate control measures. This raises the question of whether we can use members of this family of virus as a vector – or carrier – to immunise the livestock not only against the pox diseases caused by the virus itself, but perhaps other diseases as well.

The effect on sheep of this virus is manifest in the form of lumpy skin lesions. This causes severe loss of productivity, suffering to the animal, and death in a proportion of the animals. There are no methods of treating this disease and until recently there had been no method of preventing it either. However, it is possible to isolate naturally occurring attenuated forms of the virus which do not produce disease but if inoculated into sheep, goats or cattle will prevent disease. In other words, these naturally attenuated viruses can be used as a vaccine. Experimental vaccinated groups which have received the attenuated virus are fully protected from the virulent (disease-causing) virus.

It has been possible to go on and identify genomic sites in these viruses in which we can integrate other genes. For example, a gene for an enzyme has been introduced which integrates in a random fashion into different parts of the DNA of the virus. Those viruses which survive the manipulation express the enzyme which can be detected in a colour reaction. This tells us that the gene into which the enzyme gene has integrated is not essential for replication of the virus.

By doing these sorts of experiments, and with background information on the genetic structure of the virus, it has been possible to identify three sites in the genome of this virus into which you can introduce foreign genetic material without altering the ability of the virus to replicate. The next stage is to use this information to target genes to integrate into the particular sites that one has identified as being non-essential for viral replication. The technique used to achieve site-specific integration is known as 'homologous recombination'. The foreign gene for integration is contructed so that attached to its ends are pieces of DNA that correspond to the gene in the virus which is at the target site for its integration. Because of their sequence homology (complementary similarity), there is a higher probability of integration at this site through homologous recombination than at other sites within the viral genome. The genetically engineered virus can then be injected or inoculated into host cells and one can select individual colonies of viruses that have taken up the foreign gene by this recombinant process. The result is the selection of viruses which contain a foreign gene and which express the protein encoded by that foreign gene.

The first experimental example in which this approach has been used successfully with the capripox virus is rinderpest. At the turn of the century a pandemic epidemic of rinderpest in the African continent resulted in the death of millions of animals. Subsequently, the disease has been controlled by vaccination. However, there have been certain difficulties in applying the existing vaccine in the field caused in particular by the short shelf-life of the vaccine. Using the procedure described a gene of the rinderpest virus has been cloned and introduced into the capripox virus genome. If it works as effectively as the traditional vaccine, the advantage of a vaccine based on capripox is that it should have a much longer shelf-life because capripox is a very rugged virus.

Finally, the fourth approach involves the use of non-infectious particles of viruses. One might ask, given that it is possible to isolate and use individual viral proteins, why produce intact particles of particular viruses? The reason is that for many simple viruses the use of individual proteins has not proved to be successful. For example, take the case of the foot-and-mouth disease virus which is composed of three different proteins in a network structure that forms the virus particle. These three proteins are intimately associated with each other. The problem in trying to mimic using just one protein the immune response that is induced by the intact virus is that if you isolate one protein from this particular structure it adopts a different structural conformation. The challenge to vaccine production against foot-and-mouth disease virus is to produce the shell of the virus without having any genetic information within it, thereby producing something that mimics the native virus in structure but does not present any genetic information or any possibility of infection to the host.

Further studies are underway to determine whether enough of these 'shell particles' can be produced and whether we can then introduce these into animals and obtain immunity similar to that induced by the parent virus. Note that we are not introducing any foreign genetic material into the animal and therefore there should be relatively little hazard involved.

The way in which animals have been selected for various traits has taken, in general, little consideration of their disease-resistance status. Therefore we have the potential of looking back within populations of the particular animals that we are utilising for agriculture for naturally occurring resistance traits which we can then perhaps introduce, through the transgenic technologies discussed earlier or through conventional breeding, into the animals currently used for food production. One of the technological advances that will make a major contribution to this is gene mapping. What one would like to do is to obtain a map of the genome of the particular species that we are interested in so as to localise on that genome the individual genes that are responsible for conferring resistance to disease. One would then have the options either of using that information in conventional breeding programmes to select for the particular traits that are of interest, or of using transfection technologies to introduce the genes for those traits into animals already used for food production. In both instances one is merely shuffling the genetic material within the population of animals of interest. The author considers that the exploitation of naturally occurring disease-resistance genes through the use of DNA technology offers a tremendous opportunity to reduce our dependency on intensive methods of disease control.

5 Embryo Transfer and Reproductive Technologies in Farm Animals

Professor Christopher Polge

The application of reproductive technologies to the breeding of farm animals provides important opportunities for the genetic improvement of livestock. In this context I would like to start by quoting from a joint statement issued by the Royal Society of London and the United States National Academy of Sciences on 'Population growth, resource consumption and a sustainable world'. This statement reflects the shared view of the two academies that 'sustainable development implies a future in which life is improved worldwide through economic development, where local environments and the biosphere are protected, and science is mobilised to create new opportunities for human progress.'

There is already evidence of the mobilisation of science for the improvement of animal productivity in advances that have been achieved during the last few decades. An important aspect of animal improvement is the application of principles of genetic selection. One of the technologies that has contributed greatly to the worldwide implementation and dissemination of genetic improvements in several of the major domestic species is artificial insemination (AI).

ARTIFICIAL INSEMINATION

Artificial insemination was first introduced as a practical measure for the breeding of farm animals some 50 years ago in several countries. Since then its use has increased tremendously, particularly in cattle breeding. Today the majority of dairy herds are bred by AI and the technology is considered to be the greatest development in agriculture in modern times.

Advances in AI were made possible both by the development of techniques for the dilution and storage of semen and by the selection of the best sires for genetic improvement by means of progeny testing. The discovery of a method for the long-term preservation of sperm at very low temperatures has had one of the greatest impacts on the development of AI. This advance was made over 40 years ago and today the preservation of semen by deep freezing is almost universally adopted for AI in cattle breeding. It has revolutionised the livestock industry by enabling the effective transfer of valuable genetic material within and between countries and by providing a unique method for genetic conservation.

One of my most poignant memories from a visit to India some years ago was to see in many small villages an AI centre with flasks containing frozen semen. Here the generally low-producing local cattle were inseminated with semen from bulls of much higher genetic merit, some of which was imported from abroad. In consequence it has been possible to more than double milk production in these areas within the space of a few years. This has not only improved human nutrition but has provided an important source of income for impoverished farmers.

By contrast, another consequence of the widespread use of AI is that it can have a major impact on breeding structures of domestic livestock, particularly in the developing countries. In some countries, for example, indigenous breeds are becoming highly diluted or even replaced as semen from more productive and improved breeds is increasingly used for promoting higher production. It is important therefore to consider the potential detrimental effect that may result from a reduction of genetic diversity in our animal populations with the resultant need for conservation. Cryogenic storage of semen provides an ideal method for genetic conservation and it is good to see that steps in this direction are already being taken by organisations such as the Rare Breeds Survival Trust in this country and the Food and Agriculture Organisation (FAO) internationally.

EMBRYO TRANSFER

Embryo transfer was introduced much later than AI, but in several countries it has already been established in cattle breeding for about 20 years. The objectives are similar to those of AI, namely, genetic improvement, but it is potentially a technique with greater impact because it is concerned with the whole animal genome rather than just half as is the case with sperm.

At present embryo transfer is based on techniques involving superovulation to increase egg production followed by insemination to produce a number of embryos. The embryos are recovered from the uterus about one week after ovulation by non-surgical procedures and they are also transferred non-surgically to surrogate recipients. Applications so far have been relatively limited and confined mainly to producing a few more calves from pedigree dairy or beef cows. In the UK, for example, about 8–10 000 calves are produced by embryo transfer per annum. Methods have also been developed for the preservation of embryos by deep freezing. Deep freezing provides an opportunity to store embryos that may have been produced by superovulation until such time as animals at an appropriate stage of the reproductive cycle are available as recipients. There is also the obvious opportunity for export or import of embryos between countries which has a greater impact on genetic improvement than might be achieved by the use of frozen semen. An additional advantage is that embryos are probably the safest way of exporting valuable genetic material from the point of view of animal health and welfare.

Recently, embryo transfer has been combined with techniques for enabling more rapid genetic improvement in cattle breeding. These are based on schemes for multiple ovulation and embryo transfer (MOET) in nucleus herds. Several calves can be produced from each cow in a short period of time and superior bulls are then identified by the performance of their sisters. Females are assessed on their own performance and that of their sisters. This technology allows selection for a number of traits which can be equally directed towards welfare considerations as well as individual performance characteristics, such as milk production for example. MOET schemes for genetic improvement are also better adapted for use in nucleus herds in developing countries rather than progeny testing which may be impossible to implement.

IN VITRO FERTILISATION AND SURROGACY

Opportunities for enabling more rapid and extensive application of embryo transfer in cattle breeding would be possible if a much more plentiful supply of eggs or embryos could be provided than can be obtained by superovulation. To this end there has long been a considerable interest in alternative methods for using the very large pool of immature eggs that are present within the ovaries of all mammals. In recent years there has been great progress in the development of techniques for the large-scale production of cattle embryos by *in vitro* techniques. These are based on the recovery of immature eggs from the ovaries of animals and the use of techniques for *in vitro* maturation and fertilisation of the eggs, and the culture of embryos to a stage suitable for transfer.

Large numbers of embryos can now be produced by these methods and their use in cattle breeding is being developed. One of the first applications is the production of embryos from beef breeds for transfer to dairy cows to provide calves of better genetic quality than can be produced by the traditional method of cross breeding. It is important in this technology to select the most appropriate genetic make-up of the embryo so that the calf which is produced by the surrogate cow can be carried to term and born naturally. Experience to date provides evidence that the incidence of difficulties at calving is no greater as a result of the transfer of appropriate embryos than following AI.

SEX SELECTION

There is no doubt that efficiencies in animal production could be greatly enhanced if it were possible to pre-select the sex of offspring that were born so that female calves were produced for dairy herd replacements and male calves produced for beef production. For some years it has been possible to detect the sex of individual embryos by the use of specific DNA probes. To do this a small biopsy consisting of a few cells is taken from an embryo (usually at the blastocyst stage when there are more than 100 cells present) and the DNA which is extracted is then analysed with a DNA sequence specific to the Y-chromosome present only in normal males. Such methods are highly accurate, but somewhat complicated and expensive to perform.

For many years attempts have been made to predetermine the sex of offspring by the separation of sperm into two populations that contain either only sperm with an X-chromosome – resulting in female offspring – or sperm with a Y-chromosome – resulting in male offspring. The sex of the offspring could then be effectively controlled at the time of fertilisation. Many attempts to separate X- and Y-sperm on the basis of possible physical or immunological differences have generally been unsuccessful so far, but recently methods using flow cytometry have been far more successful. In most of the domestic species the X-chromosome is slightly larger than the Y-chromosome which results in a DNA mass difference of about 3–4 per cent. This difference can be detected by the use of a fluorescent dye to stain the sperm nuclei and the sperm can then be separated by cell sorting techniques on the basis of the small difference in the intensity of fluorescence.

In experiments in the USA in which rabbits were inseminated with sperm sorted by this method, significant shifts in the sex ratio of offspring produced were achieved. For example, 94 per cent of the offspring were female in does inseminated with sperm samples enriched for X-sperm and 81 per cent were males in does inseminated with Y-enriched sperm. The number of sperm that can be sorted effectively in a given time is too few to enable the use of this technology in traditional artificial insemination in cattle breeding. By contrast, sufficient numbers of sperm can be sorted to enable fertilisation to be carried out *in vitro*, and these opportunities are presently being explored.

CLONING

Greater use of improvements that are made in cattle breeding could perhaps be achieved through the production of larger numbers of embryos from animals of superior genetic merit. A single embryo sometimes splits during early embryonic development to produce genetically identical twins. Embryo splitting has been carried out in the laboratory and it has been shown to be possible to produce twins or quadruplets by these methods. The possibility of producing larger numbers of genetically identical animals has more recently been explored by using techniques of nuclear transplantation in which individual nuclei from an early embryo containing 32 or 64 cells are transferred to unfertilised eggs to create new embryos. Although some limited success has been achieved, it is clear that much more fundamental research is required before this technology can be applied. Apart from its relative inefficiency, it has become apparent that some of the calves that have been produced from embryos derived by nuclear transfer have been oversized. The reasons for this effect are not clear and further research is obviously needed.

REGULATION

There is a legitimate concern that the technologies that are used in animal breeding should be controlled effectively, not only from the point of view of animal health but also with regard to their effects on animal welfare. In the UK there has long been legislation governing AI, and with the growing use of embryo transfer, appropriate measures are being put in place to regulate the collection and transfer of bovine embryos. These procedures will be clearly governed by the veterinary profession and no collection or transfer of bovine embryos may be carried out by a person who is not a member of an approved bovine embryo transfer team or a veterinary surgeon. The main welfare provisions are that the use of epidural anaesthesia will be mandatory prior to the collection and transfer of embryos. It has also been agreed that recipient cows will be required to be examined by a veterinary surgeon prior to the transfer taking place. The veterinary surgeon must be satisfied that the recipient is fit to receive the embryo and that there is no reason to believe that the cow will not carry the calf to term and calve naturally.

These regulations should do much to ensure that the growing use of embryo transfer in cattle breeding will be appropriately controlled and used in the most effective way.

I believe that the reproductive technologies applied to animal breeding described here have an important part to play in the creation of new opportunities for human progress. We should see that they are used appropriately and with due consideration for animal welfare. In the wider context of the preservation of the biosphere, the pressures resulting from the growing population of mankind and the consequent effects on the environment are leading to the loss of many indigenous species of animals. It is therefore also important that existing and emergent reproductive and genetic technologies are used for conserving genetic diversity in wild as well as domesticated species.

Discussion I

Anon: I campaigned against BST and I have campaigned against genetic engineering. But when I look at the cows wintered indoors rather than standing outdoors in bad weather, which when left with their horns on damage each other, I do feel that genetic engineering of cows so that they would be born without horns and would not have to be dehorned by mechanical means would save a lot of suffering for the animals. I also feel that the correction of the faulty genetic make-up of human beings suffering from serious genetic diseases would be valuable. What does Jeremy Rifkin think?

Dr Jeremy Rifkin, Foundation on Economic Trends: Let me describe human problems arising from genetic disorders to put this in context. There are some 3000 recessive single gene disorders. Should we eliminate all of them? I once asked the Chairman of Molecular Biology at Harvard. He said, 'Absolutely'. He forgot basic biology. Recessive traits are essential to maintain the genetic variability of the gene pool for the species. If you narrow down the gene pool for a short-term gain you do so at the expense of the hardiness of the species to withstand environmental assaults.

We are also starting to find out that these traits perform other functions. Sickle cell anaemia's carrier state prevents malaria. We have located the cystic fibrosis gene and found out it prevents certain cancers. We have not yet understood the subtleties of all the functions that recessive traits may perform.

Having said all that there is also the question of priorities. Do we engage in radical chemotherapeutic intervention after the problem has already been caused or do we deal in preventive help? For example, we all have genetic predispositions for various diseases. We also know that the environment we create can trigger those. So how should we deal with heart disease, breast and colon cancer, strokes? Should we wait until people gorge themselves with grain-fed beef and a rich animal fat diet which triggers those predispositions and then genetically change their blueprint, or do we encourage a sophisticated, nutritional regime so that we can be healthy? To my mind we have to take a look at the alternatives in each of these fields and not believe there is just one course for medicine or agriculture or energy or pharmaceuticals.

Then there is the question of the eugenic implications of human genetics. At what point would you say 'No' to either germline or somatic gene therapy? Should we try to eliminate the gene for Tay–Sachs disease or should we wait? Should we eliminate the gene for Huntington's chorea? Should we abort a foetus that may develop Alzheimer's disease at the age of 50? When we become involved in programming our offspring we become involved in a new parent–child relationship where the child becomes a creature of architectural design. Basically and philosophically, it changes our relationship to life. Before we embark on that road we should have a very public spirited discussion and debate.

Finally there is the question of eugenics and somatic gene therapy. Scientists are now looking at how to insert a gene that would make someone resistant to a chemical carcinogen. Instead of cleaning up the chemical environment that triggers the cancer, why not just make a somatic gene change that renders people immune from the causes of cancer in the environment? Employers may well like that approach which may absolve them of the responsibility to safeguard worker health.

I could go on but each of these areas requires a thoughtfully engaged public debate because we are embarking on a journey into the 'age of biology'.

Dr Eric Millstone, Science Policy Research Unit, Sussex University: I want to contest an assumption that was shared by Kevin Ward and Christopher Polge about the basic context within which biotechnology is being developed. Although they are right to say it is very important to improve the public understanding of science, it is no less important that the understanding scientists have of social and political affairs needs to become more sophisticated. It is, I believe, a gross over-simplification to maintain that people are hungry because there is too little food. It is not correct, as I understand it, to claim the human population is growing or has been growing exponentially, or that food production is growing or has been growing more slowly than the population. Population growth has been below exponential rates consistently for the last several hundred years and, moreover, the productivity of agriculture has grown more rapidly than the population. I am not saying there is not a good case for improvements in agricultural productivity, but people starve essentially because they are too poor to buy the food that is available. If you misconstrue the problem of hunger as simply a technical problem requiring a technical solution, then you will over-simplify the possible technical solutions. That this is the case is evident from the example of the 'green revolution' which successfully improved agricultural production but, nonetheless, led to increased hunger by amplifying the inequalities and poverties within the population of most of the countries within which it was introduced.

Dr Kevin Ward, CSIRO, Australia: I certainly do not want to attempt to over-simplify the food problem, indeed, I agree it is very complicated. However, I think that the population on this planet is, in fact, growing exponentially – I disagree with you on that. But that is not the point. The point is that eventually we are going to need more food on the planet. We cannot get more food on the planet the way we are going at the moment, using natural processes, and I simply do not see any way in which we can use conventional selection to increase food to the level it is going to be needed.

Prof. Christopher Polge, Animal Biotechnology, Cambridge: I agree with Dr Ward's argument and that is why I think it is very important that we take biotechnology to the poor countries so that they can increase their own indigenous wealth. I have seen a graph which shows how world food production has kept pace with the growing population and it follows exactly the graph of the increased use of nitrogenous fertilisers. Perhaps we do not want that, and if not we should look at alternative means.

Juliet Gellatley, The Vegetarian Society: Feeding grains to animals and then eating those animals is an extremely wasteful and damaging use of land and, Dr Ward, you probably know that whilst we could feed even the present world population if everyone were vegetarian, we certainly could not do so on a western animal product-based diet. So why do you offer, as an answer to global hunger problems, feeding genetically engineered cereals to genetically engineered animals? Why do you not advocate vegetarianism or veganism?

Dr Kevin Ward: We are not vegetarian – we are in fact omnivores and we have evolved that way. The point I am trying to make is that it is extremely difficult for us to exist in an energy-restricted world purely on a vegetable diet. The species that thrive on a vegetarian diet are the ruminants which have evolved to handle the very poor quality roughage-type diet. It is possible for us to exist on something like a vegetarian diet but we need a lot of energy input to do so. My point is not to address the food problem of the planet; I concede that the food problem is largely going to be solved over the next couple of decades by redistribution. However, we are going to need animals. Surely, we do not expect that this world is going to get by without any form of animal usage at all? My own specific interest is in sheep. I think that the wool that sheep produce is going to be very important for us. If we can increase the efficiency with which sheep can produce that very important fibre wool, then we will be doing something significant for the improvement of the population that is going to be on this planet in about 20 or 50 years time.

Albrecht Muller, Germany: This is a question for Dr Ward. How can you hope that the situation of the poor people in the world will be improved by a more efficient sheep that is patented?

Dr Kevin Ward: In Australia we are not patenting sheep – let me be quite clear about that. CSIRO, the organisation that I belong to, is a national organisation and we work for the people of Australia. We envisage that the sheep that we are producing will be disseminated to user groups who will then use them for increased productivity. However, how they will be disseminated *globally*, I do not know. That is one of the reasons why I think a meeting like this is so important. There are many international issues that have to be sorted out. I do not see the answers at the moment, I see many problems and I think that some of the problems are included in your particular question.

Malcolm Eames, BUAV: Dr Ward, I took your talk to suggest that as a result of genetic manipulation, fewer sheep would die from drought. Now surely that is a misconception. The limiting factor in droughts is the availability of water. If you make the sheep more efficient in eating what forage there is, all they will do is eat it quicker and then die.

Dr Kevin Ward: My aim is actually to reduce the sheep population in Australia. We have 160 million sheep in Australia and I think that is far too many. I would dearly like to see 80 million sheep which would be possible if I can make sheep twice as efficient. I would like to see fewer sheep, but I want to see the wool production continue because wool is such a useful fabric for us. That is a social issue which I am certainly not equipped to come to grips with, but if I can provide the means whereby we can reduce the sheep population, then it is up to the social fabric of society to reduce the numbers of sheep. Now I am sure that this is not going to happen immediately.

Anon: From the contributors so far, I see two points emerging. There is no doubt that genetic engineering has the power to create great good but it also has the power to cause great harm. All the contributors have talked about openness, about being able to discuss the issues involved. But let me compare this apparent willingness to openness with the position regarding farm animals and laboratory animals at the present time. There is very little public discussion over how we treat such animals. So will the scientists who are saying they will be open about the issues that genetic engineering raises live up to their word?

Prof. Christopher Polge: It is true that millions of laboratory animals are kept in this country. Many of them are kept for toxicology testing, which is mandatory for all new products. One of the areas in which the transgenic technology we have described is potentially very helpful is to develop totally *in vitro* systems for the examination of potential harmful effects of products, for example possible mutagenic effects. A big effort is also being made into the development of immortalised cell lines that can be used in this respect, and I hope that that is appreciated [see Part IV].

Anon: Dr Morrison emphasised 'productivity', which to me simply means 'profits', but what about animal welfare?

Dr Ivan Morrison, AFRC: I think I did qualify my statements on productivity with references to animal welfare, and a number of my comments were related particularly to developing countries. People have got to realise that in some developing countries the large numbers of livestock cause erosion and, for many of the animals, the levels of mortality are incredibly high; in many areas up to 50 per cent of the calves fail to reach 6 months of age. Now under those circumstances one could envisage using some of the methodologies that I described to introduce disease-control measures – diseases are the major factor in those losses – and thereby being able to produce more food from fewer animals in an extensive system, which would be a major contribution to animal welfare.

Dr David King, Genetics Forum: I would like to make one or two comments in response to the contributions of Dr Ward, Dr Morrison and Professor Polge.

One unfortunate aspect of animal biotechnology is the use of hormones to encourage animal growth. In my opinion the presentations rather 'slid over' this issue and I think that may be because some of the animal biotechnology industry is ashamed of the rather sad failure of a lot of that research. Scientists do not yet understand the endocrine (hormone) system of animals, and when they have inserted additional growth hormone genes into pigs there have been all kinds of unpleasant side-effects for the animals. This is an example of how, despite their lack of understanding of the physiological system they wished to manipulate, scientists nonetheless proceeded with the experiment to insert additional growth hormone genes into the animals. And why did they do that? I think it is clear that it is because of the commercial pressures on the companies that are engaged in animal biotechnology to get a product on to the market that will make them money. That is a rather unfortunate aspect of animal biotechnology.

I cannot resist making a comment on the debate about population, a topic which seemed to dominate a lot of the presentations. A point that was not made, but which should be understood, is that whilst there is a population problem, that population problem is here, in the developed world. The problem is that we, as 'northern' consumers, consume too much, and that it is our over-consumption which is depleting the world's resources, not the people in the 'south'. We are the problem.

Finally, I would just like to say something about the issue of conservation of genetic resources which Professor Polge talked about. I think his contribution was a perfect example of the 'mind set' that is dominant in the scientific community, namely, that the so-called superior genetic animals should be used on farms, and the genetic material of the others should be conserved in a 'genetic museum' in case of future utility. That is a very dangerous approach. We have seen the result in vulnerability to disease that can result from that. I think we need now to start thinking more seriously about an agriculture where genetic diversity is used on the farm.

PART II

Transgenic Farm Animals: Regulation and Impact

6 The Impact of Genetic Engineering on the Welfare of Farm Animals

Professor John Webster

I wish to consider the effects of biotechnology on the welfare of farm animals, that is, on the quality of life as perceived by the animals themselves. I contend at the outset:

1. that I write as a scientist who believes that our capacity to do good is enhanced if we base our opinions on observation and reasoned analysis;

2. I do not consider farm animals to be machines simply for the service of man but sentient creatures with a quite advanced perception of quality of life which includes not only visceral sensations such as hunger, cold and pain but also some higher perceptions of pleasure, frustration and depression.

ANIMAL WELFARE

I have defined the requirements for animal welfare in terms of 'five freedoms' which seek to guarantee good food, comfort, good health, security and freedom of expression. I also argue that this analysis of animal welfare is comprehensive. However, farm animals enjoy a sixth freedom denied to us, namely freedom from introspection. Provided quality of life is met by attention to the first five, it matters little to the animal whether our intention is to love it, ride it, eat it, or sacrifice it to find a cure for cancer or to test a new cosmetic. These things are real problems but our problems not theirs. Moreover the welfare implications of artificial breeding are defined almost entirely by the consequences of the procedure not the procedure itself or who did it. Thus the broiler fowl has been crippled and the bulldog incapacitated not by biotechnology but by the application of traditional selection methods accompanied by natural mating. On the other hand, farm animals which have been genetically engineered to contain a valuable human gene such as Factor IX become very valuable creatures, and are indeed pampered like racehorses. None of these animals agonises over whether it is the subject of a patent application.

GENETIC MANIPULATION OF FARM ANIMALS

The genetic manipulation of farm animals must be distinguished from the scientific study of how genes control biological function and development. Genetic manipulation is a commercial enterprise and will only be funded in the reasonable expectation that it will one day make money for its sponsors. The things that might attract a potential sponsor are as follows:

1. Genetic manipulation of animal feedstuffs to increase nutrient yield and composition.

2. Genetic manipulation of digestion so as to increase nutrient availability.

3. Manipulation of metabolism so as to:
   (i) increase the absolute rate of meat or milk production;
   (ii) alter the composition of milk or meat (e.g. more protein, less fat).

4. Increasing reproduction rate in the breeding female.

5. Increasing the rate of genetic progress via, for example
   (i) embryo transfer, (ii) cloning.

6. Conferring genetic resistance to infectious disease.

7. Manipulation of cognition by gene deletion within the central nervous system.

8. Insertion of human genes into farm animals so that they can serve man as sources of
   (i) pharmaceuticals, (ii) cells, tissues and organs.

Let us examine each of these procedures in terms of animal welfare as defined by all six freedoms (that is, including freedom from introspection). Procedures one and two, manipulation of the supply and availability of nutrients, are unlikely to affect animal welfare directly. This seems at first sight to be a sound application of new technology. If the enthusiasm of the gene-handlers in this area needs to be constrained, it is not on the grounds of welfare.

The manipulation of animal metabolism so that they produce more meat or milk clearly has the potential to create problems. According to the Farm Animal Welfare Council (FAWC) these include, 'the manipulation of body size, shape or reproductive capacity by breeding, nutrition, hormone therapy or gene insertion in such a way as to reduce mobility, increase risk of injury, metabolic disease, skeletal or obstetric problems, perinatal mortality, or psychological distress' (FAWC 1988).

The modern broiler fowl bred by conventional means to reach slaughter weight at 42 days is as unacceptable as the Beltsville pig genetically engineered by the insertion of human growth hormone. Both animals have been crippled by being forced to outgrow their strength. Enthusiasm for the use of biotechnology as a method for increasing meat and milk production has largely evaporated. This reflects in part an awareness of welfare problems associated with accelerated growth rate (see, for example, Webster 1990), and also a greater sense of scientific and commercial realism; growth is a very complex phenomenon and involves more genes than the engineers can handle. Moreover the commercial sponsors, having burnt their fingers with bovine somatotrophin (BST), are no longer driven by the question: 'Can I produce this food more cheaply?', but by the more difficult question: 'Will the public buy it?'

The manipulation of reproductive rate by hormone therapy, gene transfer or insertion of extra embryos seemed, like the manipulation of growth rate, to be a good idea at the time when the prime motivation was increased productivity. However, in a strictly commercial sense, the gains resulting from an increased supply of meat animals relative to the overhead costs of maintaining the breeding females have to be balanced against certain losses. For example, there is clear evidence that artificially increasing the reproductive rate in both cattle and sheep is associated with an increased risk of disease and death to both mother and offspring.

New breeding techniques which involve the collection, manipulation and transfer of embryos can undoubtedly increase the rate of genetic improvement. In cattle, I can accept that the procedure, as perceived by the cow, is not that different from artificial insemination (AI). In sheep, both embryo transfer and AI with frozen semen involve laparoscopy, and this must distress the sheep somewhat.

The 1986 Animals (Scientific Procedures) Act incorporates what is, in my opinion, an outstanding improvement in ethics relative to its predecessor, the 1876 Cruelty to Animals Act. Under the new Act no procedure likely to cause any degree of pain or suffering whatsoever can be carried out until there has been a full cost–benefit analysis which poses the question: 'Is the cost to the animal in terms of suffering, however slight, justified by the likely benefit to society?' Obviously the greater the cost the greater the demand for justification. At present, farm animals do not enjoy the benefit of any law similar to the Animals (Scientific Procedures) Act 1986 for laboratory animals, and I think they should. If they did, we would be faced by questions such as: 'Are the surgical procedures involved in multiple ovulation and embryo transfer justified by an increase in the rate of genetic gain in the national dairy herd from its current niggardly 0.5 per cent per annum to perhaps 2 per cent per annum, if we are lucky?'

The use of biotechnology to impart genetic resistance to infectious and metabolic diseases in farm animals must surely be universally acknowledged as a thoroughly good idea since the costs are negligible and the benefits to both people and animals are high. I would like to think that this would be the way forward for farm animal biotechnology, but I doubt it. There is less profit to be made in freeing an animal from a disease for all time than in giving it a daily injection of BST. Of course, if the genetic engineer could patent his healthier animal he might find the exercise worth while. I am pretty sure the healthier animal would not mind, in which case neither would I.

In recent years scientists studying the genetic basis of information storage and transfer within the brain have discovered that deletion of single genes in laboratory rodents can dramatically impair the learning process and so reduce the animal's cognitive capacity to understand its surroundings and act accordingly. I have said before that I believe science *per se* to be amoral. Science is the disinterested pursuit of knowledge, and morality resides in its application, that is, how we use that knowledge. This piece of science leads us into an ethical mine field. If we are concerned about the quality of life experienced by farm animals, can we overcome that problem by destroying their perception of quality? I do not like the idea, but perhaps I am failing to live up to my own rational standards.

The final and most dramatic application of biotechnology, both in a scientific and an ethical sense, concerns the manipulation of farm animals to make them genetically more like humans. Genetic engineering of sheep so that they secrete pharmaceuticals such as Factor IX in milk has met with general approval. The benefit to man is obvious. Establishing the gene in the germ line of the sheep carried a small welfare cost to a considerable number of sheep but carries no cost to subsequent generations. Genetic manipulation of pigs so that their tissues become more compatible with man and can be used for spare-part surgery may, at first sight, seem more alarming to us, but as far as the animals are concerned, the sheep that secretes pharmaceuticals in milk or the pig that is being reared as a cardiac donor is likely to be treated with much more care than the ewe starving on a hillside or the early-weaned pig in a wire cage. Webster's fourth law states that the welfare of an animal raised for commercial purposes is directly proportional to its cash value. The question we must ask ourselves is: 'Do we consider it more or less acceptable to regard the pig as the source of a new heart or a sausage?'

THE NEED FOR REGULATION

I have considered the impact of biotechnology on farm animal welfare strictly in terms of welfare as perceived by the animals themselves, and almost without regard for our own motives. I am convinced that our motives are of no direct concern to the animal. However, I am equally convinced that we owe it to ourselves to weigh the welfare costs of any form of tinkering with animals against the potential real benefits to society. There should be open debate of these issues, they should not be carried out in secret. However, it will be in the interest of neither man nor animal if solutions to specific cases are decided solely on the basis of entrepreneurial pressure from the proponents of biotechnology and the gut reaction of the general public. I believe that the commercial application of biotechnology and artificial breeding in farm animals should be subject to law similar to that contained in the Animals (Scientific Procedures) Act 1986.

There are those who say such legislation will impede the progress of science. This argument is invalid. Science is the pursuit of new knowledge and understanding. This new law would apply to the commercial application of knowledge and ask: 'Does the end justify the means?' Transposition of this question into good law would, I believe, help to protect farm animals from unnecessary suffering, protect the consumer from products they neither want nor need, and give livestock farmers some respite from, on the one hand, exaggerated claims of the biotechnologists, and on the other, a growing sense of alienation from a public which is less and less inclined to recognise them as men and women who like working with animals and are only trying to do an honest job.

7 Transgenic Animals in the Market-place

Professor Derek Burke

Scientists are now able to inject DNA into the nucleus of a fertilised egg of an animal in the test tube where this injected DNA will be integrated into the cellular DNA of the egg, thus creating a transgenic egg. After implantation back into the animal, the transgenic egg develops in the normal way to result in birth. A small proportion of the transgenic animals born will produce the gene product as a protein which may have an effect on their growth or physiology.

Apart from the injection of the DNA, the procedures for creating transgenic animals which I have just described are similar to those involved in *in vitro* fertilisation (IVF). Thus, the only new ethical concerns arise from the effects of the injections, the integrated DNA, and the protein products of the DNA on the animal's growth and behaviour. There is also a separate question as to whether or how transgenic animals may be used in the food chain. It is this latter issue and the role of the Advisory Committee on Novel Foods and Processes (ACNFP) which is the main focus of this chapter. First, however, I should like to comment on technical aspects of the procedures for creating transgenic animals.

PROCEDURES FOR CREATING TRANSGENIC ANIMALS

The first point is that the process of transgenic manipulation is rather inefficient. Before the gene product (the protein) can be produced, the gene that is injected has to be both integrated (that is, become a stable part of the host DNA) and expressed – that is, to produce, through the intermediary function of a messenger RNA (mRNA), the appropriate protein. Both processes – integration and expression – are inefficient: only about

10 per cent of the animals born of treated eggs will contain integrated DNA, and only about 1 per cent of the animals will express the gene, and that will be at variable levels. In other words, for every 100 animals reared from an injected fertilised egg, about 90 are normal, another 9 contain a foreign gene which is not being expressed, and only one animal will produce the foreign protein in its body.

The second point is that the process of producing a productive transgenic animal is rather slow because, as I will describe, it is the lactating animal that is useful. For an animal to mature to the stage of lactation may take from a few weeks in the mouse, to a few years for pigs, sheep, goats and cattle. Only then can the animal be tested for its capacity to produce the foreign protein. The next stage is to breed from the small group of heterozygous animals that produce the foreign protein in order to create a homozygous population for continuing production.

The third point is that the process varies in its effect on the animal. The first successful experiment involved the injection of the gene for human growth hormone into mice (see Bulfield 1990). Those animals that contained an active gene grew more rapidly and to a larger size than normal mice. This is because human growth hormone is active in mice and is not immunogenic. The implication of these results is also that the production of mouse growth hormone is rate-limiting for growth. The results obtained with mice are not obtained with all other species: insertion of additional hormone may have no effect, or may indeed have a deleterious effect, for example in the the case of pigs which express human growth hormone gene (see Wheale and McNally 1990a Part 1).

THE ADVISORY COMMITTEE ON NOVEL FOODS AND PROCESSES

I am the Chairperson of the ACNFP. In 1989/1990 the ACNFP received two enquiries about issues arising from the presence of human genes in goats or sheep. The gene for human Factor IX, the blood clotting factor, had been inserted into the germline by the method I have described, together with appropriate signals upstream from the DNA to ensure that its product – Factor IX – would be secreted in the milk. In addition, the appropriate carbohydrate was added to the Factor IX by normal mechanisms found in the host animals. The milk is thus a good source of the human Factor IX, from which it can be readily purified.

Other proteins that have been produced by this method, or for which this method is proposed, are:

- human alpha-1-antitrypsin – lack of which can cause emphysema – has been produced in sheep by Pharmaceutical Proteins in Edinburgh, and in one cow in quantities of up to 35 grams per litre of milk;

- human serum albumin has been produced in goats;

- human lactoferrin, a protein involved in iron adsorption, has been produced in cows by GenPharm International in California. So far the only animal to express the gene is a bull, which illustrates another technical hazard of the technology!

The above examples are *replacement* proteins, that is, relatively large-volume products with existing markets, where the larger amounts produced by transgenic animals give this technology a clear advantage over cell culture. A dairy cow can produce 300 kg of milk protein per year, and up to 10 per cent of this can be the foreign protein.

In addition, transgenic farm animals may be used to produce *new* pharmaceuticals. Human tissue plasminogen activator (t-PA), a protein which dissolves blood clots, has already been produced by Genzyme, a company based in Cambridge, Massachusetts, USA. Moreover, Genzyme has proposed the use of transgenic farm animals to produce human cystic fibrosis transmembrane conductance regulator – the protein that is faulty in cystic fibrosis.

The question put to the ACNFP was whether the normal animals – that is, the 90 per cent where there had been no integration of the foreign gene and, of course, no foreign gene expression – could be released into the food chain? It is self-evident that it would be wasteful to incinerate, or otherwise destroy, these animals, and that if the technology were to become widely used, then the numbers of these animals could be quite large.

We considered that the animals concerned could be divided into three broad classes. The first class comprises the one per cent that are producing the foreign protein in question, and there is currently no intention to let them enter the food chain. The second class are the 10 per cent that contain the foreign gene but are not expressing it. It could be argued that these are normal animals that happen to contain a stretch of nucleotides that, in another species, codes for a protein, but in these animals is no more than 'junk' DNA – non-coding, apparently non-functional DNA, of which there is a lot in all animal genomes. However, the release of such animals into the food chain *is* a matter of public concern, as was shown by an incident in which some pigs containing a non-expressing gene were released into the food chain in Australia without consultation or notification, and a public storm ensued.

Finally, there are the normal animals, which only differ from other animals fertilised *in vitro* in that a needle has been poked into them. It is, of course, useful to be able to distinguish between these two classes. The polymerase chain reaction is of such exquisite sensitivity that the presence or absence of the new gene – which, if present, would be in all the cells of the animals – can be determined with great confidence. Of course, if absent the gene could never appear in subsequent generations.

So the technical evaluation is clear: this 90 per cent of transgenically manipulated animals are normal by all the criteria we can use, and the conclusion seems self-evident – they could be eaten or used for breeding stock. The remaining 9 per cent contain some foreign DNA – but only foreign in that the nucleotides are arranged in a particular order not found in the host, and not foreign in any intrinsic sense. And, of course, we are eating foreign DNA all the time – not only the DNA from animals and plants from our conventional diet, but also from that of our changing diet. Every time we try a new species of fruit from the supermarket we are encountering new 'foreign' DNA.

Earlier this year the USA Agriculture Department's Food and Safety Inspection Service gave a Texas company permission to slaughter six cattle for human consumption after gaining assurances that an attempted gene transfer experiment two and a half years before had been unsuccessful. We on the ACNFP are aware that a technical evaluation may not be enough to meet the public's concerns. In addition to these technical factors, we will use other criteria such as:

- Risk – Why do we have to increase any risk to life or the environment? We want zero risk.

- Dread – A concern about the peril that might ensue if the technology went wrong, as, for example, nuclear power.

- Stigma – A negative association with a whole technology, for example, food irradiation.

- Outrage – How dare they do that to us? Surely it's unnatural? Why weren't we told?

These are real issues, and the ACNFP is attempting to deal with them, since unless they are dealt with the public may not accept the new technology, and I believe that would be a loss. The way forward must involve better communication of risk and benefit, increased transparency of the approval process, and listening to, and trying to address, the concerns of the public.

So although we were clear that those animals which contain no integrated gene could be released into the food chain on technical grounds, we alerted the Minister at the Ministry of Agriculture, Fisheries and Food (MAFF) to the wider concerns. As a result, the Government appointed an independent expert study group to report on any ethical concerns which may arise from the supply of foods produced by genetic modification (MAFF 1992).

The study group's aim will be to consider likely future trends in the production of genetically modified food and the concerns which might arise from their consumption, and to recommend how those concerns might be addressed. It will be chaired by the Reverend John Polkinghorne, President of Queen's College, Cambridge. Other members are: myself, Professor Derek Burke, Chairman of the ACNFP and Vice-Chancellor of the University of East Anglia; Professor Brian Heap, Director of Research at the Agriculture and Food Research Council (AFRC) Institute of Animal Physiology and Genetic Research; Mrs Harriet Kimbell, Vice-Chairman of the Consumer Panel; Mr Richard Callaghan, Technical Director of ASDA plc; and the Reverend Dr Arthur Peacocke, Warden-

Emeritus of the Society of Ordained Scientists and Honorary Chaplain of Christ Church Cathedral, Oxford.

It should be emphasised that the ethics study group's remit does not extend to the ethics of genetic modification as such, but is limited to the food use of organisms from genetic modifications. In 1992 the group issued a consultation paper to nearly 200 organisations and institutions seeking their views on concerns which could arise, in particular on the following four areas:

1. The transfer of human genes to food animals. Examples include the sheep in which a human blood clotting factor gene had been incorporated. Both modified animals and those in which incorporation of the foreign DNA had not occurred could potentially be used for food.

2. The transfer of genes from animals the consumption of whose flesh is forbidden to certain religious groups, for example, pigs for Muslims and Jews, or cattle for Hindus, to animals which they normally eat. In the USA cattle have been modified to carry the porcine (pig) growth hormone gene.

3. The introduction of animal genes into food crops, the consumption of which may be of particular concern to vegan vegetarians.

4. The use, as animal feed, of organisms containing human genes. For example, yeast can be modified to produce human proteins of pharmaceutical value; consideration might be given to disposing of the spent yeast as animal feed.

At present none of the products referred to above is entering the human or animal food chains but it is the ethical concerns that could arise from such a practice that the ethics study group aims to address.

Initially the study will attempt to assess likely future developments in this area. A range of views have been invited on the ethical concerns that could arise from the type of developments described above, and also on the following points:

- Whether there should be controls on the food use of any particular type of product, and if so, what form such controls should take and which products should be covered.

- Whether concern mights be met through labelling or other forms of information to consumers, or by some other means.

The responses will, unless explicitly precluded, be made publicly available after the ethics study group reports to the Minister at the end of 1993.

8 Acceptance or Rejection? A Consumer View on Genetically Engineered Animal Products

Dr Tim Lang

Manufacturers and scientists are asking people to consume the products of genetic engineering. If it is not you or I who is to consume, it has to be someone else. Most of the questions I want to ask in this chapter regarding the consumption of genetically engineered animals apply equally to other food commodities. The special questions raised by genetically engineered animals, from a consumer point of view, merely intensify the need for answers to questions already being tabled by the consumer movement.

THE QUESTIONS

The questions I want to ask of genetic engineering are these:

- What are the consequences of eating the products of genetic engineering?
- What difference would better information make?
- If there is to be education, who is to do it?
- What interests are at stake?
- Who has responsibility to act on any concerns?

Although it is nigh impossible to lump all consumers together on anything, genetic engineering is performing a near miracle, uniting disparate movements, differing philosophies, and different interests. Almost everyone is troubled – not panicking, not hysterical – but troubled. And the more people find out about genetic engineering, the more concerned they get. Please let me stress that point and I must explain why. The notion most giant companies – and remember most companies investing in genetic engineering are hardly your local family food manufacturing companies – used to have of 'the consumer' was appallingly arrogant and patronising. The consumer's role was to buy and consume, and to feel grateful for the food industry's miraculous capacity to fill supermarket shelves and to produce thousands of new food products each year to tempt our taste buds.

Happily, the battles over food quality and safety we had in this country in the 1980s have terminated that patronising (or matronising) approach. Both industry and government were shaken by the vehemence of public outrage that emerged following the food scandals in the UK in the 1980s. A free market government, used to praising the food industry as a paragon of what the private sector could do, was reminded of the lessons learned by its predecessors in the mid-nineteenth century: that state regulation is essential if the public is to have confidence in its food.

In an industrialised society, where few people live close to the land or the sources of their food, a confidence factor cannot be supplied by market forces. The issue is not whether to regulate, but *how* to regulate sensibly, flexibly and openly. The 1980s food scandals culminated in the Food Safety Act 1990 and an apparently more consumer-friendly approach within the Ministry of Agriculture, Fisheries and Food (MAFF) and the food industry – and about time too, thought the British public.

Even before the extraordinary rows of the 1980s, we in the consumer movement with a special interest in food policy had a different approach with regard to the introduction of new food technologies than that which used to be taken by the food industry on the whole. Whatever we think about free market economics, our assumption is that the market (of whatever variety) only works if the consumer is informed. So our message is: tell the public and let them weigh up the arguments and make their own minds up.

Alas, it seems that the proponents of genetic engineering have yet to learn this important principle. We can quickly understand why. The investment is huge: billions of dollars have been invested in the genetic engineering industry. An estimated US$0.5 billion went into developing just one product, genetically engineered (recombinant) bovine somatotrophin (rBST) (*New York Times* 1990). Such investment forces the food industry to have to sell its products – whether they are a genetically modified meat or potato.

No wonder the four multinational drug companies [Monsanto, Eli Lilly, Upjohn, and American Cyanamid] which have invested in genetically engineered BST have received such a drubbing over this product. Their attempts to get rBST licensed in the late 1980s and to begin recouping their investment came to the attention of a consumer already highly sensitised to vested interests in food. If cows (which are herbivores) could develop a new disease – bovine spongiform encephalopathy (BSE) (or mad cow disease) – from being unwittingly fed feed containing the remains of diseased sheep, then clearly the moral standards of industry could sink very low. The rhetoric of 'consumer sovereignty' turned out to mean deliberately keeping the consumer – whether human or bovine – in ignorance.

It is this consumer scepticism that now challenges the might of those who wish us to eat or drink the products of genetic engineering. Will the genetic engineers rise to this challenge?

THE SURVEYS

These days I get a survey through the post or a letter requesting an interview from an underpaid academic about once a month. Public opinion and what public relations (PR) mandarins like to call 'opinion formers' (that is, potential sources of anything ranging from media comment to organising genteel trouble) are now features on genetic engineering companies' maps.

I welcome such interest in public opinion, even though I must confess to slight weariness at my opinion constantly being solicited. (I have made my views clear endlessly – or is the hidden agenda to tie people like me up in endless form-filling!) Alas, the interest in public opinion is a little belated. I have a distinct feeling that the PR contingent is interested in trying to close the stable door after the horse has bolted.

In 1991 the MAFF published a survey on food labelling (MAFF 1991a; see also FAC 1990). It showed that around a quarter (23 per cent) of consumers would use labelling information telling them that genetically manipulated organisms (GMOs) were contained in the food. Of these, 28 per cent said they were against genetic engineering, 15 per cent said they just needed to know, and 12 per cent had health worries. In 1992 a survey for *Supermarketing* magazine found that this figure had gone up: 85 per cent wanted their retailers to label genetically engineered food and only 9 per cent disagreed (*Supermarketing* 1992). The demand for the right to know has grown.

You could interpret this increase as the result of wild and woolly luddites successfully selling consumers their view of genetic engineering. Whilst such an interpretation might be appealing to the paranoics within the industry, alas it is not true. The same survey showed that 48 per cent would still consume genetically engineered products, and 32 per cent (an increase on MAFF's figures a year earlier) said they would not. My interpretation – born not just from controversy over genetic engineering but from consumer debates about many technological processes such as irradiation, additives and pesticides – is that once people are given more knowledge they get more concerned about:

- the speed of introduction of novel products;
- the lack of consultation; and
- the imbalance of power between giant corporations anxious to recoup their investments and the individual consumer, who is required (not forced) ultimately to consume these products (or else there is no sale).

This imbalance of power between producers and consumers fuels concern over civil liberties and rights, and mention these and companies get decidedly nervous, smelling legal costs and that bane of marketing: public opinion!

The market model is premised, as I have noted, upon consumer access to information, and it is the theme of information that I shall now pursue.

INFORMATION

Guidelines for the labelling of foods produced using GMOs were set by the Food Advisory Committee (FAC) after advice from the Advisory Committee on Novel Foods and Processes (ACNFP). The FAC guidelines say that there is no need to label a food containing a GMO if the GMO is nature-identical and the gene transfer is intraspecific. The justification for this category of exemptions is that in these cases genetic modification techniques have merely accelerated the effects of traditional breeding practices which could produce similar results over a longer period of time (MAFF 1991b). This exemption from labelling is tantamount to giving a green light to industry to produce only nature-identical products. The biotechnology industry however was not happy with the labelling term – 'product of gene technology' – which was suggested for the two categories for which the FAC wanted labelling – novel food products of GMOs and foods from trans-species GMOs. Industry was concerned that this phrase would unduly worry consumers.

Now we learn that the ACNFP is scrutinising a genetically engineered interesterified fat for use in infant feed formulae as a breast-milk fat substitute (ACNFP 1992). The public outcry, notably from organisations promoting breast-feeding which have had a bitter battle with infant food companies which undermine breast-feeding by mothers through pressure-selling infant food formulae, underlines public concern about the onward march of genetic engineering. Concern is not always about the product *per se*, such as a breast-milk substitute, but the process behind it.

Frankly, the moral position taken by the four multinational companies over rBST, when they hid behind the UK's Medicines Act 1968 and allowed unlabelled BST-derived milk to go into the general milk supply, was dreadful. No wonder there is now what we might call a battle for hearts and minds over genetic engineering.

The genetic engineering revolution is one of the great social shifts in science and technology. We are moving from an industrial culture to a bio-industrial culture. At home, watching the television, reading the papers, talking to family, friends and workmates, people are aware that the speed of change is rapid and potentially threatening. For these reasons, I would like to see all genetically engineered food products clearly and simply labelled.

If people are to be asked to eat GMOs, whether animal- or plant-derived, they have the right to know that they are eating GMOs. In the long run, after decades or centuries of experience, perhaps labelling will be irrelevant. Today it is paramount, which is why, we presume, the corporate interests and their friends in government resist the idea so furiously.

I am not surprised at this resistance to labelling, but if it continues then the whole theory of the free market collapses. I repeat, for markets to work there must be sufficient numbers of producers to give consumers a choice, and there must be information and education among consumers. If information is deliberately withheld – as it is being at present – the market is exposed as a sham. Far from 'getting the nanny state off our backs', big business is exposed as wanting the nanny state to act on its behalf, keeping consumers in ignorance but doing what business wants.

EDUCATION

A senior UK government committee, the Advisory Committee on Science and Technology (ACOST), asked several years ago that there be a concerted educational programme about genetic engineering (ACOST 1990, paragraphs vii and xv). They wrote: 'The power of public feeling must not be underestimated; consumer resistance and fears for safety and pollution, for example, can seriously encumber commercial prospects' (ACOST 1990, p. 23).

Even if only motivated to bolster 'commercial prospects', this appeal for public information was at least public spirited. However, we received almost nothing. The industry's (and Government's) failure to respond to what is, after all, a fairly reasonable request has occurred at a time when public debate in all Northern rich economies is focused on education and curricula, unlike any time since the 1960s. Do we have to conclude that the absence of effort for public education on genetic engineering is not incompetence?

The education task has thus been left to a small but honourable number of voluntary organisations – the Athene

Trust (and its sister, Compassion in World Farming (CIWF)), Genetics Forum (GF), the Food Commission (*Food Magazine* 1992), and Parents For Safe Food (PSF). Unfortunately, the debate that ensues often takes on the form of a David *versus* Goliath battle, but at least the information percolates out, through the media as well as through these small public pressure groups' own publications.

DECISION-MAKING

The question most often put to consumer representatives is: 'Haven't you got this out of proportion? Surely', the questioner continues, 'it is not in companies' interests to pull the wool over consumers' eyes or to sell products with an even slight risk attached.' To this we invariably say: 'Fair point'. No one is accusing the companies of a conspiracy; the process of obfuscation is more complex, more messy and on occasions, 'dirtier'. What is at stake here is the issue of trust. Confidence in one's food is important but confidence is also a very personal and fragile psychological construct.

The experience of the consumer movement of scientific decisions is a bit like the citizen's experience of justice. There can be no satisfaction without representation, and the only good decision is one which is not only just, but seen to be just. So the committees which sit in our name are important. MAFF set up an expert ethics group to report on ethical concerns arising from consumption of genetically engineered foods. Set up in September 1992, it was been given a year to report to the Minister (MAFF 1992). There was one consumer representative – the Vice-Chair of the Consumers Association – among six in the study group [see also Chapter 7].

One out of six people may seem a small victory, but it is significant, as anyone with experience of Whitehall could tell you. After its drubbing over the food scandals, MAFF is trying to appear to be, and to be more, consumer-friendly. There is now a consumer panel which meets the junior MAFF Minister regularly. Consumer representatives also sit on the key FAC. There is a trades unionist and an environmentalist on the Department of the Environment's Advisory Committee on Releases to the Environment (ACRE). At the time of writing, one consumer has been appointed to the key ACNFP; and another to the new MAFF Group on ethical concerns arising from genetically engineered food.

This is surely some progress. When we review the international scene, however, this rug of small victories is neatly pulled from under our feet, as I explain later.

CHOICE? WHAT CHOICE? AND WHO BENEFITS?

Choice is at the heart of the consumer society's core beliefs. We all like to think we have choice. Social scientists show that we humans are both remarkably varied and remarkably limited in the range of our behaviour. Modern consumption is in fact schizophrenic – part private, part public; part individual behaviour, part mass behaviour.

The world is narrowing. We are fed by giant companies whose turnover dwarfs many a nation state's gross national product (GNP). As we meet, the world stands poised for a new world economic order. The General Agreement on Tariffs and Trade (GATT) is an international treaty which seeks to remove barriers to trade. The rationale for the GATT might be trade liberalisation, but the reality looks more like deregulated commerce in the interest of powerful business, rather than the consumer.

This new treaty has huge implications for genetic engineering (Lang 1992). Nonetheless, when the (then) European Community (EC) threatened to impose some controls on biotechnology companies, many of which are USA-based, the companies threatened to withdraw their operations from Europe entirely (see Haerlin 1990). Under this threat, the EC created a regime so favourable to biotechnology industries that in the talks to renegotiate the GATT, companies in these industries complained that what the GATT proposes is not as favourable to their interests as the climate for investment created within the EC!

SUSTAINABLE DEVELOPMENT AND FREE TRADE

The world's 5.4 billion people are divided into three consuming classes (Durning 1992). The billion, including you and I, who over-consume, who raid the world's larder, who warp the statistics; three billion in the middle who consume more or less at sustainable levels; and one billion at the bottom who underconsume. Ecologists have taught us that the post World War II boom in the rich countries of the 'North' cannot continue.

Bearing this in mind, I urge consideration of the following three issues: first, the theory and practice of 'free markets'; secondly, the bias in the adjudication system enshrined in the GATT to arbitrate on scientific judgements; and thirdly, the minutiae of the intellectual property rights section of the GATT.

The Theory of Free Trade

Free market theory argues that human wealth is best gained by removing barriers to trade and allowing each country to produce what it is best suited to do. 'Today', say the free traders, 'the world needs to dismantle national barriers designed to protect home industries. If this were to happen, the three billion of the world's population in the middle consuming class would raise their standards of living to emulate our rich patterns of consumption, and benefits would trickle down to the one billion underclass.'

The above theory sounds great, but unfortunately it does not hold up. Alan Durning's (1992) excellent *World Watch Institute* report shows that the world cannot sustain the excessive consumption of the rich one billion; the ecology of the world cannot accommodate such energy waste and feed itself. A more decentralised, less intensive system of food production is urgently needed. New models for trade and development that give priority to the environment, health and sustainable development are required.

In some respects, the GATT talks represent the ultimate hijack. Today's trading system is being warped by rich developed trading blocks – the USA, Europe and Japan – when they should consume less. The struggle over genetic engineering lies at the heart of this matter, for the model of genetic manipulation we are currently being 'sold' is a high-tech vision determined by a relatively small number of powerful companies.

Arbitrating Disputes

The GATT's proposals for the use of science in the trade dispute settlement process are elegant and clear (GATT

1991). A United Nations body, the Codex Alimentarius Commission (known as Codex), will act as arbiter on matters of scientific judgement and thus 'influence' the decisions.

The GATT process would work as follows: if Sweden or Botswana wanted to ban the sale or consumption of GMOs on ethical or scientific grounds, any company or country objecting that this would keep out their products could make a complaint to the GATT secretariat and invoke the GATT disputes procedure. Three wise men (chosen from a list kept by the GATT secretariat in its Geneva headquarters) would make up a panel, which would seek advice on the scientific basis for the ban from Codex.

The process of making scientific standards on matters of public safety and health is of critical public interest. The GATT proposals for resolving trade disputes mean that whoever sets international standards will play a powerful role. Codex is supposed to be a committee of governments, but its participants are in fact about one-third drawn from industry (Avery *et al.* 1993). Non-governmental organisations (NGOs), representing the interests of animal welfare and environment, are absent. (On a rare day you can sight an occasional consumer.) Thus an increasing number of NGOs now believe that Codex is wholly inappropriate for the sensitive task the new GATT assigns to it.

A New Patents Regime

The question at the heart of today's debate on genetic engineering is: 'Who owns a gene?' What is at stake here is the privatising of the global commons. The GATT proposes a new regime for arbitrating on this question. My reading of the GATT proposals – not just in the intellectual property rights (IPR) sections, but also those on technical barriers to trade and sanitary and phytosanitary standards – is that a tidal wave is about to hit us.

A study for the EC has concluded that about 90 per cent of all primitive cultivars and landraces used in the 'North' derive from the South (Jenkins 1991). Meanwhile we in the 'North' – in the name of development – have destroyed or threaten much of our own biodiversity. The Biodiversity Convention agreed in Rio de Janeiro at the 'Earth Summit' has, frankly, not tackled this ethical issue; indeed it could be interpreted as having accelerated and accepted the process of the commodification of nature (Jenkins 1992), and this worries me.

The GATT's IPR proposals are explicit: 'Patents shall be available for any inventions, whether products or processes, in all fields of technology, provided that they are new, involve an inventive step and are capable of industrial applications' (GATT 1991, Article 27, paragraph 1). Furthermore, Part iii, Section 1, Article 41 of GATT's draft Final Act states that any country signing the GATT would have to change their laws to: 'Permit effective action against any infringement of intellectual property rights covered by this agreement' (GATT 1991, Part iii, Section 1, Article 41, p. 76).

You would probably agree that the above sounds favourable towards industry. Yet the large biotechnology companies were unhappy about the GATT's draft IPR agreement, arguing that it is not permissive enough. They point to the biotechnology policy in the EC as being far more favourable to the growth of the biotechnology sector than the GATT.

Third world governments want control over their genetic resources; multinational companies desire to have the right to exploit that genetic material. Third world countries want negotiation over intellectual property rights; corporations want legal rights. Thus the conflict between the two will continue. NGOs and the public must wake up to this struggle for control, and I believe that we must ensure that the GATT as it currently stands does not have the last word.

We in Europe have been so fixated on the EC's proposals for the patenting of life forms that we have paid less attention to a far more important new global system of patents or intellectual property rights. The GATT's IPR proposals will mean that what was public property can become private property. Under the GATT, the global commons can be taken into private ownership and become a traded commodity (Shiva 1992).

THE CHALLENGE FOR THE FUTURE

The contributors to this book bear witness to the rising conflicts of interests over genetic engineering. These interests include those of the consumer movement, whose campaigning groups are up against some powerful forces. However, although the interests involved are formidable and the stakes are high we have a number of significant levers we can pull.

Most significantly, the companies need us, the public, to

consume their new products. Ultimately, they know that we can flex our muscles at the point of consumption. Already, for instance, a high-profile resistance movement has been initiated by some chefs in the USA who refuse to sell genetically modified products [see ch. 2]. So far the products at stake are vegetables such as the 'FlavrSavr' tomato (*Food Magazine* 1993 p. 3). I foresee more such resistance, and when the products are genetically modified food animals, the moral grounds of the producers will be all the more tenuous.

Who can tell what lies ahead? I see a number of avenues emerging including: educational efforts; a moratorium on genetic engineering developments; better organisation of NGO cooperation; and action after the new GATT arrangements are agreed.

We need to force the government's hand on education by redoubling – trebling – our effort on education. If they will not take the issues to the public, sensibly and mutually, then NGOs should. Our duty as citizens, not just as consumers, is to ensure that debate and information goes as wide as possible. While private discussions with individual companies, industry umbrella groups and representatives of the state sector are welcome, our primary responsibility is to take the issues to the public. We do not want to emulate the industry by privatising discussion when we believe in the public's right to know.

Controls are urgently needed. I favour a moratorium on development. To the more paranoid in the genetic engineering industries, this might seem 'luddite', a term redolent with historical analogies. It is wrong to portray the luddites as having been against new technologies (Hobsbawm and Rude 1973). Just as they, in their time, were not anti-technology, neither are we. The issue today is not just whether humanity wants or needs the products of this technological revolution, but also who is promoting them? On what terms? As defined by whom? In whose benefit? A moratorium would allow breathing space on both sides of the debate. The products are being introduced too fast, with too little public understanding, with next to no consumer information, and driven by the financial need of investors to recoup their costs.

We need to organise ourselves better. Already the small NGOs concerned about genetic engineering are better linked than they were and are sharing information. In the UK an important cross-sectoral NGO committee, the Committee on Biotechnology and Food (COBAF), was set up in 1992 to further this process. Within Europe, the Patent Concern coalition brings together environmentalists, consumers and animal welfarists, as did the BST campaign in the late 1980s.

Such alliances are beginning to anticipate rather than just respond to new products and processes in genetic engineering. Fire prevention may not hit the headlines as much, but it is always preferable to fire fighting. We need more such alliances, not just Europe-wide, but globally. We have put together a 185-member Consumers Network on Trade to monitor how trade will affect issues such as genetic engineering.

Under the new GATT (as per the draft of December 1991), the right of national governments to resist unwelcome developments in the name of protecting their consumers is seriously weakened. With committees as biased as Codex to adjudicate, then Heaven help us, for in my view, Codex should be abolished or at least radically reformed. We need to learn about and plan for the implications of deregulation of trade – both within the European Union after the signing of the Maastricht Treaty, and globally following the signing of the GATT.

9 A Critical View of the Genetic Engineering of Farm Animals

Joyce D'Silva

Advocates of transgenic farm animals often claim that genetic engineering is really not any different from the selective breeding that has been going on for centuries. What modern genetic engineers do is simply faster and more accurate and therefore of enormous potential benefit to the farmer. Some even claim that by using transgenic techniques in combination with cloning they may soon be able to produce super-herds of identical, high-performance animals – a boon to farmer and supermarket alike. However, as I explain below, there are flaws at every stage of the argument that genetically engineered animals will be a boon to the farmer and the food processing industry. But before undertaking this explanation, I shall examine the adverse animal welfare effects of traditional selective breeding.

TRADITIONAL SELECTIVE BREEDING AND ANIMAL WELFARE

In traditional breeding methods genes cannot be exchanged between unrelated species. However, if we look at the results of selective breeding, it can been seen that it is far from being a harmless technique. Selective breeding has a long history indeed, but refinements in the last 30 years have probably resulted in greater changes to farm animal physiology than in the previous 200 years. Moreover, the results have been far from innocuous for the animals concerned. Let us examine some of these animals in order that by seeing what selective breeding has achieved we can perhaps deduce what genetic engineering may achieve in the decades to come.

Take the modern broiler chicken which is bred for meat. These birds now grow to slaughter-weight in just 6 weeks, about half the time it took 30 years ago, and that 6 weeks is being reduced by about one day a year. Presumably there must be a stopping point somewhere around the corner – but we do not know where! As for the chickens themselves, the rapid growth rate has led to severe physiological problems.

The Agriculture and Food Research Council (AFRC) estimates that up to 80 per cent of broilers suffer from leg problems, ranging from mild deformity to incapacitation (Blindell 1992). It seems muscle (that is, meat) development has been at the expense of skeletal strength. Some deformed and incapacitated chickens, unable to walk even to the feed and drink provided, sink down onto the floor to endure a slow death unless put out of their misery by an alert and merciful stockperson. Lame chicks spend more time resting on blisters and hock burns caused by the ammonia which builds up in the soiled litter underfoot.

Broilers are also increasingly succumbing to congestive heart failure because their cardiovascular system simply cannot keep up with their rapid growth rate. So, much of the chicken's short life may be spent in discomfort, pain or even agony. So much for the achievements of the selective breeding of broilers!

Let us now look at the modern turkey – a creature domesticated comparatively recently. It is nowadays virtually impossible for male turkeys to mount females as they have been bred to develop huge, meaty and profitable breasts, which make mounting impossible. Instead the breeder males are 'milked' every few days and the females are artificially inseminated. It has been said that this is probably the only accepted form of bestiality!

But is it not gross to have developed such creatures at all? Pigs also suffer an increasing degree of leg problems, lameness and heart failure due, at least in part, to their faster than ever growth rates and quicker, heavier muscle development.

The modern high-yielding dairy cow produces ten times as much milk as her calf would have needed, had it been left to suckle from her. As a result, there has been an increase in udder injuries and diseases parallel with the increase in milk production (Ekesbo 1992). Currently, over a third of dairy cows in the EU suffer from painful mastitis every year.

So to claim that traditional breeding techniques are accepted and acceptable is just not true. They may have become accepted in farming circles – sadly even in many veterinary circles – but they are not considered acceptable by those whose prime concern is the welfare of the animals.

TECHNICAL MERIT OF TRANSGENIC FARM ANIMALS

Let us look now at the claim that genetic engineering is far more accurate than traditional breeding. Theoretically, the possibility may exist for accuracy to be achieved; occasionally, it is achieved; but for every 'bull's eye' there are a myriad of failures.

One of the chief areas of research activity has been with growth hormone genes to increase growth rates and to act as muscle promoters, because lean meat is more desirable and profitable than fat. But whilst adding an extra growth hormone gene to an animal may make it grow faster and leaner, it may have other effects as well.

Dr Rexroad of the United States Department of Agriculture (USDA) has written how his transgenic lambs with added growth hormone developed a lethal form of diabetes (Rexroad *et al.* 1990). The lambs described by Dr Ward (see Chapter 3) showed degeneration of the liver and kidneys at necropsy which may have reflected diabetes-associated degenerative changes, and Rexroad *et al.* (1990) conclude, not surprisingly, that early death of the transgenic lambs prevented development of lines of transgenic sheep. Similar results were obtained by Nancarrow and others, including Dr Ward: of 12 transgenic animals expressing the growth hormone gene, all died before one year of age, with clear evidence of damage to liver, kidney and cardiac functions along with abnormal plasma levels of glucose and insulin indicating a diabetic condition (Nancarrow *et al.* 1991).

Let us continue to let the researchers speak for themselves. Groups led by both Bolt and Pursel, for example, have been experimenting with added growth hormone genes in pigs at the USDA's facility at Beltsville (see Bolt *et al.* 1988; Pursel *et al.* 1989). Accuracy is hardly the correct adjective to describe the results here. After the eggs have been microinjected with several hundred copies of the human or bovine growth hormone gene, they are transferred into the oviducts of recipient females. On average, fewer than one pig per litter will be transgenic. Of these transgenic animals only 60 per cent will actually express the foreign gene.

The researchers are frank about the 'bad news'. They report the existence of several health problems in transgenic pigs due to the excessively high levels of growth hormone. For example, pigs expressing high levels of growth hormone tend to be lethargic, they exhibit indications of muscle weakness, and some are susceptible to stress. Others tend to lack coordination in their gait, probably because their feet are rather tender. So far, all of Bolt's group's transgenic gilts that express the growth hormone transgene have been anoestrus* and their reproductive tracts are infantile. Boars also tend to lack libido, but can be used for breeding with the help of electroejaculation and artificial insemination. A number of transgenic pigs have died from gastric ulcers before they reached one year of age. Others have had lesions on the stomach lining when slaughtered for carcase evaluation. Some of the pigs show evidence of arthritis and the authors remark frankly that to date the technology has not produced a single animal with a beneficial transgene (Bolt *et al.* 1988).

Vernon Pursel's group, also at the USDA's facility at Beltsville, records the problems encountered with their transgenic pigs. Having examined, killed and carried out necropsies on both transgenic pigs and control pigs, Pursel reports that the most common clinical signs of disease associated with transgenic expression include lethargy, lameness, uncoordinated gait, exopthalmos (bulging eyeballs) and thickened skin (Pursel *et al.* 1989). Some of the transgenic pigs also showed other severe health problems ranging from gastric ulceration, severe joint inflammation and degenerative joint disease to heart problems and pneumonia. Interestingly, Pursel notes that centuries of selection for growth and body consumption may limit the ability of the pig to respond to growth hormone (Pursel *et al.* 1989).

In 1991, at the Conference of the British Society for Animal Production, Dr John Clark of the AFRC station at Edinburgh is reported as declaring that in experiments conducted in the UK in 1990, 11 399 pig eggs were injected with a foreign gene, yet just 67 transgenic animals were produced, and in sheep, of

---

* anoestrus is the non-breeding period or period of absence of sexual urge

4500 injected eggs only 34 transgenic animals resulted (Clark 1991). Pursel reports that in 3 years of gene transfer studies in pigs, only 8 per cent of the 7000 injected eggs developed to birth and about 7 per cent of those born were transgenic – an efficiency of 0.6 per cent (Pursel *et al.* 1989).

Thus, to date, genetic engineering is a 'hit and miss' technology and, sadly, many of the misses have disastrous effects on the animals concerned. Such experiments look likely to continue, and indeed, increase in number in the years to come.

CLONING FARM ANIMALS

Scientists working on transgenic animals are looking for a certain kind of genetic perfection – the right promoter, the right gene for this or that trait, the perfect combination – the grade A result. Once achieved, the next step would be to clone – to produce identical copies of – 'genetically perfect' transgenic animals by transferring gene-containing cell nuclei from such an animal into egg cells. An example of the results of such manipulations could be a herd of genetically identical cows. Let us examine the results of research in this area, as reported by the scientists themselves.

Dr N.L. First of the University of Wisconsin, for example, speaks of an era of exciting possibilities for rapidly propagating and tailoring animals to meet product and environmental demands (First 1990). Dr First (and he is not alone in this) also speculates on the use of cloning combined with genetic engineering and embryo transfer to achieve what he describes as well-organised systems for widescale production, sale and transfer of genetically superior animals, all tailor-made to supermarket specifications regarding carcase quality, fat coverage and joint size. Everything tailor-made to consumer preference! Genetic engineering optimists view the production of identical superior animals as the ideal answer to the requirements of the supermarket. Professor Peter Street of Reading University proclaims: 'In effect, with this implanted, designed embryo, if we then are able to manipulate the feeding system, we can design the whole carcase, if you like, from embryo to plate to meet a particular market niche' (Street 1992). One senior researcher in this area has told me that he believes every farmer's ideal is to have his own brand of animal and a contract with Marks & Spencers!

But are there not inherent dangers in cloning? Already, researchers have found that calf foetuses produced by nuclear transfer have abnormal growth rates. The cows giving birth to these extra-large calves are therefore more likely to experience difficulties, require more caesarean sections and as a result, more of the cloned calves die (Bondioli 1992). Those scientists who have made further clones from the initial clones, that is, mutiple-generation cloning, have found lower pregnancy rates and higher abortion rates. So far, only third-generation calves have been achieved (Stice 1992). Furthermore, the other inherent danger in cloning is that the production of genetically identical animals means that such animals are not only identically super-fast growing, super-lean, *et cetera*, but also identically vulnerable to the same pathogens. Thus, one strain of disease to which all cloned animals were highly vulnerable could 'wipe out' the entire herd. (In an analogous way, this has already happened with crops.)

Were agriculture to become dependent on cloned animals, then there would, of course, also be a significant loss of genetic diversity, which could spell long-term disaster for the livestock industry.

DISEASE-RESISTANT FARM ANIMALS

I have so far explored genetic engineering for enhanced productivity, but research is also being undertaken to develop farm animals with inherent resistance to diseases which result in, for example, massive vaccination programmes for poultry and the huge prophylactic and therapeutic use of antibiotics. To reduce dependence on vaccination and antibiotics would probably be preferable both to the animal and for consumer perceptions of animal products and probably to the welfare of the consumer of animal foods.

An overall criticism of this work is that most of the diseases being worked on are those that are the endemic diseases of the factory farm. There is a real danger inherent in this work: by genetically engineering these animals to be resistant to the endemic diseases of the factory farm, are we not condemning them to a continued existence in those same factory farm conditions – conditions often of filth and overcrowding and the frustation of natural instincts and physiological needs?

Donald Salter, of Michigan State University, has written of his ambition regarding the chicken which is to produce a transgenic 'super chicken' using the new genetic engineering techniques. This 'super chicken' would have germline inserted genes that would convey resistance to a variety of different pathogens that affect the 'livability' and productivity of chickens (Salter 1988).

For Duane Kramer and Joe Templeton, of Texas A&M University, the *raison d'etre* for their work on disease resistance in mammals is that animal health problems account for around US$14.5 billion in losses per year (Kramer and Templeton 1988). They admit however, that there are no publications documenting the generation of a disease-resistant cow, pig or sheep through recombinant DNA technology (Kramer and Templeton 1988). These researchers introduced interferon genes into cattle, but the two successful pregnancies ended in terminations. They warn that some genes will be detrimental to early development and will have to be used with regulators that will delay expression of the gene until the desired time. They point out that we still know very little about why animals are susceptible or resistant to various diseases (Kramer and Templeton 1988). In other words, this work is still at the 'shot in the dark' stage and it may be some time before disease-resistant farm animals are found on our farms. In the interim, the agripharmaceutical industry can heave a sigh of relief.

MOLECULAR 'PHARMING'

Another important development in transgenic animal engineering is 'molecular' pharming – the production of useful proteins in the milk, blood or eggs of animals. This has been forecast to be the most economically important aspect of biotechnology implementation for animal agriculture. Dr Floyd Schanbacher of the Ohio Agricultural Research and Development Center points out that the market saturation for milk and eggs would make alternative markets for these products especially attractive (Schanbacher 1988). He speculates that were the technique to prove successful in cows so that a cow could produce 3 grams of a highly valuable recombinant protein a day, her daily production would have a value of over US$200,000. He concludes that to him it is obvious that a strong commercial incentive is driving the development of the technology (Schanbacher 1988). He foresees that, if successful, this technology will allow dairy cows to be used not just to produce high-value pharmaceutical proteins, but also bulk recombinant proteins of lower value to be used as nutritional supplements or to produce milk specifically high in certain food processing qualities. Otto Postma, of Gene Pharming Europe, declares the cow to be an extremely suitable production medium for making these proteins, with no safety or technical problems (Postma 1989).

What are the drawbacks of these developments? There is still technical uncertainty about the integration of the gene and its level of expression. Trials with dairy cows are inevitably costly and slow as it is 2 years before another generation can be observed. This means it will take time for a viable herd of such transgenic animals to be developed. Dr Schanbacher forecasts that the development of the technology will require the sacrifice or biopsy of significant numbers of dairy animals (Schanbacher 1988). In addition, it is recognised that changing just one component of milk, particularly the proteins, may have dramatic effects on the nature of the milk which could render it unsuitable for human consumption, or indeed consumption by future calves.

Producing human proteins in the blood of animals has been shown to be possible, but has inherent disadvantages. For example, not too much blood must be extracted otherwise the animal will die. Also, the presence of the products of foreign genes in the animals' bloodstream could have severe biological effects on the animals themselves.

Producing human blood-clotting factors in the milk of sheep or cows could guarantee a product free from human immunodeficiency virus (HIV) or hepatitis agents. However, it would also be necessary to exclude other infectious agents which could be present in the milk.

In theory, if foreign genes can successfully be targeted to the mammary gland they should carry less risk to the welfare of the host animal as the mammary glands are relatively isolated from the rest of the body's system. In fact, however, some welfare problems have arisen. For example, researchers at the USDA's facility in Beltsville have found that high expression of the murine whey acidity protein in the mammary glands of transgenic pigs had adverse effects on the physiology of the udder (Pursel *et al.* 1989).

This work is still in its early stages both as regards the successful expression of economically viable levels of the recombinant protein in the milk, and as regards the extraction and purification of the proteins from the milk. Some experiments in molecular pharming have shown that the end-product has significant differences in structure to a similar protein produced in cells in the laboratory. Pursel *et al.* (1987) has forecast problems if the human protein has to be separated from its animal equivalent.

Proponents of molecular pharming triumphantly proclaim that it can only be good for the animals concerned. Of course, creatures as valuable as these will be looked after well. But if the techniques are successful and commercial production takes off, I find it hard to envisage that transgenic sheep, goats or cows will be allowed to roam the fields. Surely they are far more likely to be kept in the most sterile conditions possible, surrounded by stainless steel, and well away from possible sources of infection such as bedding material, or indeed other sheep, goats or cows. So we will have a type of hygienically superior 'factory farm' developed in the laboratory.

Of course, if developments like the 'DNX project' to produce human haemoglobin in pigs' blood get off the ground, then we know where we are as regards the welfare of the transgenic animals. Projections state that if 100 000 pigs were to be killed and desanguinated yearly, they would yield US$300 000 million worth of human haemoglobin (Hodgson 1992).

I think it is also important to note that producing high-value pharmaceutical proteins via the mammary glands of transgenic animals is not the only way to obtain these products. Tissue culture of insect or mammalian cells can be used and is already being used. In the future it may be possible to produce such proteins in plants – surely a far better idea?

Although potentially the most benign of the genetic engineering technologies now being applied to farm animals, molecular pharming is obviously not without its problems. At the moment it appears to be greatly overrated by its proponents who seem to see it as a cure-all solution for many human deficiency conditions. Whether or not this proves to be the case is not known at the present time.

Philosophically speaking, I find it hard to be at ease with

the description of such animals as 'bioreactors' (see Wright *et al.* 1991). Truthfully, we are all bioreactors of a sort, but we know we are much more than that; so too, I believe, are farm animals.

EMBRYO TRANSFER IN FARM ANIMALS

I would like to make a few comments about embryo transfer as this technique is integral to most transgenic work because once the gene manipulation has taken place, the resulting embryos have to be implanted into recipient animals.

To produce large numbers of eggs for this transgenic work, female animals may be injected repeatedly with hormones to induce super-ovulation. Of itself this is stressful to the animal. In some instances, of course, eggs are recovered from the ovaries of slaughtered heifers at the abattoir. This at least relieves the pressure on living animals. The new technique of locating ovarian follicles by the use of ultrasound and removing fertilised eggs from the ovaries of living animals has now been developed. This may sound innocuous, but is not so if it means animals have to be interfered with on a weekly or bi-weekly basis to procure a sufficient number of eggs. The Institute for Animal Production in the Netherlands estimates that each cow should yield 900–1000 ova (eggs) of which 15–200 will be fertilised. Egg collection can can take bizarre turns. Doctor Bob Stubbings of Guelph University has extracted eggs from 7 month-old calf foetuses and fertilised them in his laboratory. He is planning to remove one ovary from each new-born calf to increase his egg collection. In his opinion, 'Cows get along fine with only one ovary' (Stubbings 1990).

The resulting implantation of the embryo may cause distress and discomfort to the constrained cow, hopefully anaesthetised by an epidural. The partially or fully anaesthetised sheep, goat or pig will endure surgery of varying degrees of severity.

In this case, animals are being viewed routinely, not as bioreactors, but as reproductive machines. It seems strange to me to see veterinary surgeons involved in this sort of work at all. How do routine hormonal injections, physical interferences and surgical operations, all for non-therapeutic purposes, fit in with the veterinary oath, an oath which is taken by all new veterinary surgeons in the UK?

GENETICS UNBOUND

Several people appear to oppose animal genetic engineering because they fear the techniques could also be used on humans for nefarious ends, for example, positive eugenics. This is a possibility, but it is not an idea I wish to pursue. Far more sinister in the short term is the possibility that farm animals may be developed that are so different physiologically or psychologically from animals as we now know them, that they will be unrecognisable. The wingless chicken is not a figment of the imagination. It would after all fit so neatly into an even smaller battery cage! And we all know battery hens cannot fly in their cages. Even worse is the prospect voiced by many, including Professor John Owen of Bangor University College, who speaks of the tremendous scope for breeding docility in species such as pigs and poultry which we want to keep in conditions that go against their natural instincts (Owen 1992). Even philosophers like Bernard Rollin of Colorado State University seem to find this acceptable. He asserts that while it is wrong to cage a burrowing animal so that it cannot burrow, there is nothing wrong in principle in changing its nature so that burrowing no longer matters to it (Rollin 1988). This seems to me to be, at the very least, warped thinking. If related to the possibility of similar work with humans we may see why we feel an intrinsic revulsion to changing a creature's nature.

So what are we facing? An enormous effort to overcome animals' inbuilt genetic constraints on growth and productivity. As if the world's health was not already being seriously compromised by over-indulgence in animal flesh and fats. A huge effort to produce tailor-made high-yielding and probably identical creatures. A major effort to use animals as bioreactors to produce pharmaceuticals and other proteins.

Surely what we, as human beings, have not faced up to in all of these endeavours is our relationship with the rest of the animal kingdom. It seems to me to be a totally anthropocentric view rooted in the medievalism of Saint Thomas Aquinas who said: 'By divine providence animals are intended for Man's use.' When divine providence is not being used to justify the genetic engineering of animals, then some evolutionary ethic of 'the strongest calls the tune' or 'might is right' seems to obtain. But is such an uncivilised viewpoint not merely fascism operating under the new name of 'speciesism'? Modern genetic engineering is heralded as the technology of the twenty-first century, a technology with enormous potential, yet so far it is being used to engineer transgenic farm animals in ways which reflect the medieval mind-set rather than that of modern liberalism.

In conclusion, it seems to me that we have forgotton that these animals are each and every one individual sentient creatures capable of experiencing a state of well-being but also capable of suffering. As long as we view them solely as commodities or bioreactors we are indeed being blind to their wholeness and to their individual being.

10 Mad Dogs Or Jackasses? The European Rabies Eradication Programme

Ruth McNally

RABIES: A BRIEF INTRODUCTION

Rabies was the acquired immune deficiency syndrome (AIDS) of a former era. Called the 'incurable wound' in medieval times, it has a fatality rate of almost 100 per cent: to date, there are only three known cases of survival among humans who showed signs of the disease. More than 1100 rabies-related human deaths were reported to the World Health Organisation (WHO) in 1983, a figure which almost certainly represents only a fraction of the actual cases (Wiktor *et al.* 1988).

It is a disease of the nervous system caused by infection with a virus, which in humans is almost always secondary to an animal bite – human-to-human transmission has never been documented. The final paralytic phase which leads ultimately to death is preceded by a phase of aggressive behaviour, irritability, viciousness and hyperreactivity to external stimuli, resulting in the characteristic fear of water.

Rabies has had a relatively high political profile in the European Community (EC) in recent years. One reason is the Channel tunnel project which has provoked, in the words of Christopher Jackson, Member of the European Parliament (MEP) for Kent East, 'nightmare visions of foreign hordes of rabid rats, bats and foxes scuttling down the Channel tunnel to spread pestilence and horrid death throughout the sceptred isle' (Jackson 1992).

Another reason why rabies has been the focus of attention is because it impinges upon the practices and ideals of the Single European Act 1986 (SEA), which endeavoured to establish four fundamental freedoms – the freedom of movement of goods, services, people and capital – throughout the EC by the end of 1992. Rabies poses a problem for the SEA on two counts. First, there is public concern that pets accompanying their owners as they 'move freely' throughout the EC will spread rabies from the four Member States with rabies – France, Germany, Belgium and Luxembourg – to the other eight rabies-free Member States. Another concern is that rabies poses a threat to the completion of the internal market. Directive 90/425/EEC establishes the principle of the abolition of veterinary checks at frontiers for trade in live animals. The existence of rabies in some Member States could restrict the free movement of livestock, thus constituting a barrier to trade within the EC which, it is maintained, could jeopardise the profitability of European stockfarming compared to its competitors.

There have been several responses to such concerns. With regard to the transmission of rabies through Channel-hopping wildlife, barrages and traps are proposed for the Channel tunnel. For pets travelling from one rabies-free Member State to another, a system of vaccination with 'foolproof' certification is proposed in order that they may 'move freely' without the necessity of quarantine in the country of destination (Wilson 1992).

With respect to livestock, there is a two-fold approach. One approach is the generalised use in Europe of a WHO-endorsed inactivated rabies vaccine as an alternative to quarantine. The other approach is to endeavour to create a rabies-free Europe. On 24 July 1989 a Council Decision (89/455/EEC) was adopted to introduce Community measures for the control of rabies with a view to its eradication or prevention using vaccines for the oral immunisation of foxes, and it is under this Decision that Community financial aid was requested for an aerial drop of 300 000 baits containing a genetically engineered fox rabies vaccine in Belgium in 1992.

RABIES CONTROL

Rabies affects dogs, wolves, foxes, coyotes, jackals, cats, bobcats, lions, mongooses, skunks, badgers, martens, bats, monkeys and humans. In Asia, Africa and parts of Latin America, the major reservoir for human rabies is dogs and cats. In countries where rabies has been the subject of control programmes, wild animals are the main reservoir of the rabies virus, and in such countries most human, pet and livestock rabies cases are secondary to bites by wild animals.

Quarantine and the vaccination of pet dogs have been successful in eliminating rabies from Great Britain, Australia and Japan, and in reducing the numbers of rabid dogs and cats in the USA where in 1984 the majority (37 per cent) of rabies cases were in skunks (2082), followed by raccoons (1820) and bats (1038), whilst numbers of cases in dogs and cats were 97 and 140 respectively (Wiktor *et al.* 1988). In Western European countries the main rabies reservoir is the red fox (*Vulpes vulpes*) which accounts for between 70 and 80 per cent of all reported cases. This makes the fox population responsible for maintaining rabies in Europe, and it is the population of wild foxes which is the target of measures to create a rabies-free Europe by 1995.

One approach to control rabies in foxes is to cull them. Crude estimates suggest that a minimum of one and a quarter million foxes are killed annually in rabies control programmes in Europe (Zimen 1980). Another approach is the random oral vaccination of wild foxes. The objective of this approach is to make a sufficient percentage of the fox population immune to the rabies virus so that its spread will be reduced, or that it may even be eradicated altogether.

Since 1978 there have been a number of programmes in Europe for the random vaccination of wild fox populations with live rabies vaccines. The earliest open-field trials in Switzerland and France used an attenuated (weakened) strain of the rabies virus called SAD (Street Alabama Dufferin). In 1983, Germany began field trials using SAD B19, a derivative of the Swiss vaccine strain, and further field trials were undertaken in other European countries. From 1983 onwards, bait containing the vaccine and intended for consumption by wild foxes was dropped from the air. The idea is that foxes that eat the bait would be exposed to the vaccine virus and so produce antibodies which would immunise them against subsequent infection with the virulent (wild-type) rabies virus. Indeed a consistent reduction in the number of cases of rabies in the regions of the vaccination campaigns was reported (see Flamand *et al.* 1992).

SAD and SAD B19 are live attenuated viral vaccines which means they are living mutants of the rabies viruses with reduced virulence by comparison with the wild-type strain. As can be imagined, although intended for foxes, the vaccine-impregnated bait was eaten by other animals. Voles, mice and other small rodents, hedgehogs, badgers, wild boars and birds – in particular pine martens – all compete with foxes for the bait. Problems with SAD and SAD B19 are that they may mutate, reverting to virulence, and although claimed to be innocuous to foxes, they are pathogenic for rodents (Brochier *et al.* 1991).

Another problem with SAD B19 is that France delayed its participation in the EC rabies-eradication plan for some years by refusing to use the German vaccine (Jackson 1992). Given that France is one of the Member States affected by fox rabies, its initial reluctance to use SAD B19 was considered to be an obstacle to the eradication of fox rabies from Europe, and thus a barrier to free trade.

In contrast with SAD and SAD B19, two new rabies vaccines (SAG1 and V-RG) were claimed to be innocuous for non-target species (Flamand *et al.* 1990). SAG1 is a rabies virus mutant of SAD, isolated by a French/Swiss/Canadian collaboration; V-RG is a genetically engineered vaccinia virus developed by the USA Wistar Institute, the Laboratoire de Virologie in Strasbourg and Transgene, and manufactured by Rhone Merieux. In the following sections I shall focus on the genetically engineered vaccinia virus – V-RG.

GENETICALLY ENGINEERED RABIES VACCINE

V-RG is a genetically engineered hybrid virus made by inserting a gene from rabies virus (ERA strain) into the genome of vaccinia virus (Copenhagen strain). The rabies gene which has been inserted encodes rabies glycoprotein G. The site for its insertion is vaccinia thymidine kinase gene. The resultant recombinant vaccinia virus V-RG expresses rabies glycoprotein G and is claimed to be less pathogenic than wild-type vaccinia Copenhagen strain. The idea is that when a fox is infected with V-RG, vaccinia virus will replicate inside the fox and express rabies glycoprotein G, thus stimulating the fox to develop immunity against rabies virus.

The first field trials of V-RG were undertaken in Belgium in 1987 and 1988. Since then it has been tested in campaigns in Belgium in 1989/90 and 1992, and in France in 1989/90 in programmes funded by the European Commission under the Biotechnology Action Programme (BAP) and Biotechnology Research for Innovation, Development and Growth in Europe (BRIDGE).

The V-RG vaccine is encapsulated in a plastic phial which is enclosed in an edible bait. In the first field trial, the bait was distributed by hand. In subsequent trials the vaccine-containing bait has been dropped by helicopter at a density of between 15 and 30 baits per square kilometre (km^2) over sites up to 42 000 km^2 in area. It is important to remember that the vaccine being tested in these trials is a living, genetically engineered virus. Between 1987 and 1992 an estimated 882 450 baits were dropped in Belgium and France, each containing 10^8 $TCID_{50}$ (tissue culture infectious dose) of the recombinant virus. That makes these experiments, both in geographical area and numbers of organisms, among the largest authorised open-field releases of genetically engineered organisms anywhere in the world.

There is no control in these field trials over the fate of the virus once it has been distributed. Any animal – including wild animals, livestock, pets and humans – which is in the area where the bait is dropped can come into contact with and eat the bait – a bait which contains a living genetically engineered virus which may replicate inside them. These trials thus engender the risk that such animals will be harmed by the virus, that they will spread it to other sites – especially in the case of birds – and that it will genetically recombine with genetic material within them to produce new mutant viruses.

The title of the proposals for the field trials of genetically engineered rabies vaccine funded under BAP and BRIDGE is 'Assessment of environmental impact from the use of live recombinant virus vaccines.' Next I critically examine the environmental impact assessment, focusing on the campaigns using V-RG conducted in Belgium between 1987 and 1990.

EFFICACY

Efficacy of V-RG was measured in pre-release studies by giving foxes bait containing the vaccine and then testing blood samples for the presence of anti-rabies antibodies. Those found to be 'seropositive' – that is their blood samples contained antirabies antibodies – were then exposed to rabies virus in order to test their resistance to infection and disease. On the basis of pre-release studies, V-RG was claimed to be a highly effective vaccine for immunising captive foxes (Brochier *et al.* 1991).

There is a formula for estimating the proportion of a target population (p) which must be effectively immunised to eradicate the infection (Anderson 1986). It is as follows:

$$p > 1 - K_T / K$$

where K_T is the density necessary to maintain the endemic persistence of rabies and K is the density of foxes in the absence of rabies. For fox rabies, K_T is estimated to be roughly 0.4 foxes per km^2 which means that more than 0.4 foxes per km^2 must be effectively immunised to eradicate rabies. The density of foxes (K) in the rural area under study was estimated to be 2 per km^2 (Brochier *et al.* 1991), therefore the proportion of the fox population that must be effectively vaccinated is at least 0.8, or 80 per cent.

In pre-release studies, in which bait uptake was 100 per cent, 83 per cent of the foxes seroconverted and of these 89 per cent were resistant to rabies (Pastoret *et al.* 1992). This means that even if bait uptake were 100 per cent, that is if *all* the foxes in the target area ate the bait (clearly an unrealistic expectation), the level of effective immunity could only be expected to be 74 per cent, which is less than that required to eradicate fox rabies. Thus, even before the trials were conducted, the data from pre-release studies did not predict that the vaccine would be effective in eradicating fox rabies in wild populations.

When tested in the field, only 30/59 (51 per cent) foxes tested in the 1989/90 V-RG campaign in France were seropositive for neutralising antibodies (Flamand *et al.* 1990) which makes the effectiveness of the vaccine questionable. Similarly, in field trials in the USA of V-RG against rabies in raccoons, only 45 per cent showed immunity in Virginia, fewer than 40 per cent in New Jersey, and in Pennsylvania the level of herd immunity was only 20 per cent. Not only are such levels of herd immunity too low to eradicate the disease, scientists are of the opinion that vaccination rates this low will not even be able to prevent its spread (*Gene Exchange* 1994).

In fact, after the 1987 and 1988 Belgian field trials the researchers abandoned the use of serological testing as a measure of vaccine efficacy in the subsequent Belgian field trials in 1989/90 on the basis that the foxes recovered from the field were dead when they were examined at postmortem, making serological testing for the presence of antibodies difficult (Pastoret *et al.* 1992). They advocated instead the use of two proxy measures to assess the efficacy of V-RG: the rate of bait uptake and the number of rabies cases in the treated area.

The first proxy measure assumes that all foxes that have eaten the bait become immunised, an assumption which would tend to inflate the estimated efficacy of the vaccine. Indeed, of 17 foxes found to have eaten the bait in the 1989/90 field trials, 6 were also rabid! This result could be interpreted as either a weakness in the method of using uptake of bait as a proxy for immunisation, or a demonstration of the failure of the vaccine to immunise. However, the researchers surmise that, because vaccination had been effective in pre-release studies, the rabid foxes must have been incubating rabies at the time of vaccination (Brochier *et al.* 1991). If field trial results are discounted in favour of pre-release results, what is the justification of doing open-field releases at all?

The level of bait uptake amongst the sample of 188 foxes (which is only 0.04 per cent of the estimated fox population in the test zone) collected in the 1989/90 Belgian field trials was 71 per cent (Brochier *et al.* 1991) which, even assuming the reliability of bait uptake as a proxy for effective immunisation, is lower than that required to eradicate fox rabies. Another factor is that the level of bait uptake amongst juvenile foxes was found to be only 49 per cent. This was believed to be because young foxes would be expected to be confined to the immediate surroundings of the breeding den and thus less likely to ingest randomly distributed baits. The low level of bait uptake amongst juvenile foxes has serious implications for the efficacy of the vaccine. This is because of the high fecundity rate of the fox, which has an average birth rate of 4–5 cubs per year, and a short life expectancy of only one-and-a-half to two-and-a-half years. This means that a high proportion of fox populations are young animals. Consequently, the maintenance of high levels of herd immunity would require repeated (preferably annual) cohort immunisation with the aim of exposing young foxes to inoculated baits at as young an age as is practically possible (Anderson 1986).

The researchers argue that V-RG has been successful in

reducing the incidence of fox rabies as indicated by the decline in the number of notified cases of rabies in cattle and sheep in the 1989/90 target area in Belgium. However, these data should be considered in the light of other information. First, it would appear that the Luxembourg province of Belgium has been subjected to field testing by both SAD B19 and V-RG rabies vaccines in recent years which would make it difficult to separate the effects of one vaccine from the other. Secondly, the numbers of foxes culled in other rabies control programmes is not indicated in the data presented by the researchers. Given that the test zones were chosen because of their high incidence of fox rabies, it is unlikely that culling of foxes to control the disease ceased altogether in these zones throughout the test period. Finally, the rabies epizootic in Belgium is cyclical with a mean periodicity of 4 years. It is not clear from the data presented that the cyclical nature of the rabies epizootic has been discounted in presenting the trends.

PATHOGENICITY

In a paper prepared for a conference presentation, researchers from the University of Liege assert that 'the modified virus is safe both for the target and relevant non-target species' and that 'no human risk is associated with the use of the modified virus' (Pastoret *et al.* 1988). However, an analysis of the risks of V-RG by the Department of Agriculture in the USA (USDA) found that the safety of the experimental rabies vaccine in humans is unknown (*Gene Exchange* 1994). Whilst postmortem examination of several wild animal species exposed to V-RG did not reveal signs of pox lesions, virologists have warned that, 'it would be unwise to draw conclusions about human virulence from observations on experimental animals' (Dumbell 1985 p. 12).

Vaccinia is a virus which was used extensively by the WHO as a vaccine in its campaign to eradicate smallpox. Whilst the WHO smallpox eradication programme is considered to have been a success, it was not without negative side-effects, including fatalities, which are instructive regarding the risks of using vaccinia as a vector for other vaccines.

Vaccinia is a virus of unknown origin, not found naturally, which is maintained in vaccine institutes and research laboratories (Baxby 1985). It is a minor human pathogen, and complications it caused when used as a vaccine against smallpox included: a generalised rash; accidental infection with vaccinia which could lead to impaired vision; eczema; and neurological complications including encephalopathy, encephalitis or encephalomyelitis among very young subjects. In addition, there were further complications in immunodeficient patients including 'progressive vaccinia', a progressive rash which sometimes resulted in death. For example, between 1951 and 1960 in England and Wales 5 000 000 vaccinations were reported with eight cases of progressive vaccinia, seven of whom died (Kaplan 1989). Whilst these incidences are not high, one may reasonably assert that there are now many more people throughout the world with compromised immune systems than there were even 10 years ago. In addition to the risk of progressive vaccinia, a relationship has apparently been observed between vaccination with live vaccinia virus and subsequent AIDS in a previously asymptomatic subject (Redfield *et al.* 1987).

The above description illustrates that whilst vaccinia virus might be harmless to most individuals, it is a pathogen for some others and may even be lethal in its effects on susceptible individuals. Human health experts have argued that whilst the risks of vaccinia were accepted when smallpox was a problem, the eradication of smallpox removed the need for, and is a strong contraindication to, any further immunisation with vaccinia virus (Dumbell 1985). This recommendation sits uneasily with the use of vaccinia as a vector, not just for rabies vaccine, but for numerous genetically engineered recombinant vaccines (see for example, Esposito and Murphy 1989), including a vaccine against AIDS developed and (infamously) tested by Daniel Zagury in Zaire.

CONTAINMENT

Pastoret and his colleagues (1988) claim that 'the genetic engineering of the virus has not changed its species specificity'. However, given the broad host range of vaccinia virus, which includes humans, rodents and the major livestock species such as cattle, sheep, pigs, camels, and buffaloes, this is not a very reassuring claim.

The researchers claim that 'the modified virus is poorly transmitted from an animal to the other in a given ecosystem and does not spread' (Pastoret *et al.* 1988). However, data from the use of vaccinia in other contexts does not support this assertion. During the smallpox vaccination campaign, it became clear that vaccinia virus was transmitted from vaccinated humans to non-vaccinated humans. Moreover, vaccinia virus has in the past spread from vaccinated humans to other animals, for example, dairy cows from where it has spread within the herd. Once animals are infected with vaccinia there is a risk that infection may spread from them to humans. This was illustrated by an outbreak in El Salvador in which about 22 farmworkers were infected (Baxby 1985).

The working definition of vaccinia is that it is not found naturally. Whether vaccinia can become established in nature is under debate. In India the virus responsible for a problematic 'buffalopox' epizootic closely resembles vaccinia virus, and some believe that it may be a variant of vaccinia virus used for human smallpox vaccination which has become established in buffaloes. If this is the case, then it demonstrates that vaccinia is not only transmissible from human to non-human hosts, but that it can become established in nature.

The above instances of the transmissibility of vaccinia virus are warnings that recombinant vaccinia virus may also be transmissible. Indeed transmission of the recombinant vaccinia has already allegedly occurred in 1986 when the genetically engineered rabies vaccine was illegally tested on cattle in Argentina (see Wheale and McNally 1988 chapter 7). The vaccine in this case was also V-RG – vaccinia genetically engineered to contain a gene from rabies virus. It was subsequently asserted that some of the 17 people exposed to the inoculated animals developed antibodies against rabies virus indicating that they had been infected with the V-RG.

The transmission of vaccinia from host to host and from species to species is an uncontrolled and uncontrollable event. This makes the selection of vaccinia, which has a broad host range, for large-scale random vaccination programmes of wild animals highly questionable. Whilst fox rabies vaccine trials have been taking place in Europe, there have been field trials in the USA of a genetically engineered vaccine against raccoon rabies. Once more, the vaccine is V-RG – vaccinia virus containing a gene from rabies virus. Ecologists in the USA have questioned the wisdom of using vaccinia, whose principal host is humans, as the basis for vaccines intended for non-human animals. One suggestion is that, in order to maximise the probability of containment within the target species, genetically engineered vaccines should be based on viruses with as limited a host range as possible. For example, vaccines intended to protect raccoons should be based on raccoonpox virus which is believed to be specific for raccoons.

Could it be that commercial considerations are overriding considerations of safety in the selection of vaccinia as the vector for genetically engineered rabies vaccines? Wistar and Rhone Merieux were involved in the bovine rabies vaccination trials in Argentina, and they, together with Transgene and the University of Liege, were involved in both the fox rabies and raccoon rabies vaccine field trials, and in each case the vaccine has been V-RG.

GENETIC STABILITY

The researchers claim that 'the modified virus is genetically stable' (Pastoret *et al.* 1988), which means that it should be resistant to spontaneous genetic mutation or genetic recombination with another virus or with the genetic material of the host.

Genetic recombination is a process in which one piece of DNA or RNA exchanges parts with another. The more similar the two genetic molecules are, the more likely they are to recombine with each other. Not surprisingly, genomes (the genetic molecules) of related viruses resemble each other in parts. Vaccinia is a member of the orthopoxvirus genus, other members of which are variola (which causes smallpox in humans), cowpox, monkeypox, camelpox, raccoonpox, taterapox and ectromelia. Although the principal orthopoxviruses have been given the status of 'species', no species-specific neutralising antibodies have been found (Dumbell 1985). The structure and replication of all orthopoxviruses is essentially the same, there is extensive cross-hybridisation (a measure of similarity) between DNA fragments from the outer genome regions of the various 'species', and orthopoxviruses hybridise with each other readily in the laboratory (Baxby *et al.* 1986).

The host range of vaccinia includes humans, cows, buffaloes, pigs, camels, rabbits, elephants, monkeys, sheep and rodents which overlaps with the host ranges of cowpox, monkeypox, camelpox, variola, and ectromelia, with each of which it shares homology. V-RG could be eaten by, or transmitted to, animals already harbouring a virus with which it shares homology. If this were to happen, genetic hybridisation between the viruses might ensue creating a new hybrid virus. The new virus created through the genetic recombination of V-RG could spread undetected through wildlife or cross into domestic species. A particular risk is posed by cowpox virus. Contrary to its name, the most commonly reported host of cowpox is the domestic cat. It may also be maintained in small wild mammals where, because it is not pathogenic, it is undetected. V-RG could recombine with cowpox virus in small mammals which eat the vaccine bait. Subsequently, cats, with their predatory nature and close human contact, could introduce the V-RG–cowpox hybrid virus into human populations (Baxby *et al.* 1986).

In addition to homology with other orthopoxviruses, the genetically engineered vaccine V-RG contains the rabies virus glycoprotein G gene which will share homology with the glycoprotein G genes of wild-type rabies viruses. Rabies viruses infect dogs, wolves, foxes, coyotes, jackals, cats, bobcats, lions, mongooses, skunks, badgers, martens, bats, monkeys and humans. If V-RG bait is eaten by a animal which harbours the rabies virus – for example a fox – the two viruses could recombine.

Who can predict the properties of a recombinant V-RG? What would its host range be? Would it be pathogenic? Would it spread to other organisms? To other species? The unanswerable nature of these questions starkly illustrates the futility of pre-release risk assessment in the face of mutation. What we do know from recent experiences – Hong Kong flu virus, the AIDS virus, seal 'influenza' virus, bovine spongiform encephalopathy (BSE) – is that some newly arisen viruses have lethal effects, can spread both within and between species, and can infect humans.

Five years after the first release of V-RG in Europe, researchers at the 1992 BRIDGE meeting on biosafety expressed the need to undertake further research to establish the rate of recombination between poxviruses. This might be a step-by-step approach to risk assessment, but clearly the steps being taken are in the reverse order (see Wheale and McNally 1990b, 1993)! What can be done about the viruses that have already been released if it is demonstrated that V-RG is highly mutagenic? Is there any indication that in the short-term at least there will be a moratorium on releasing V-RG until the risk-assessment studies on its potential for homologous recombination have been completed and indicate that the risk is extremely low? This looks unlikely: in 1993 Rhone Merieux was seeking a product license for V-RG in the USA and the EU (see McNally 1994, 1995).

RISK–BENEFIT ANALYSIS

Some might argue that environmental risks are worth taking when considered in relation to predicted benefit. Brochier *et al.* (1991) claim to have investigated the economics of the V-RG vaccine-bait dispersal programme, which they describe as follows:

> The average yearly cost of rabies in Belgium (1980–89), including treatment of humans, animal diagnosis, compensation to farmers for the culling of infected livestock, and the culling of wild foxes, is estimated to be 400 000 ECUs/km^2, or 88 000 ECUs per annum for the area under study. These figures do not include the cost of vaccination of domestic animals nor the salaries of civil servants. In comparison we estimate the overall expenditure during the three campaigns [November 1989, April 1990 and October 1990] of vaccine bait distribution (bait, helicopter and personnel costs) to be 118 000 ECUs. Because vaccination following eradication can, in principle, be interrupted or subsequently limited to the borders of the vaccinated zone, long-term maintenance of a rabies-free area by peripheral vaccination with live recombinant vaccinia virus may well be economically justifiable. (Brochier *et al.* 1991 p. 522)

Thus by their own account, which omits both the cost of developing the vaccine and the costs of the potential pathogenic 'side-effects' of the vaccine described above, the economic benefit of using V-RG is a potential *future* benefit, which is dependent upon its efficacy in the field. Yet, as I have argued earlier, there is a paucity of data on the efficacy of the vaccine, and the data that is available suggests that it may not achieve the levels of herd immunity necessary to eradicate rabies from wild fox populations.

However, let us assume that fox rabies could, as Brochier *et al.* (1991) suggest, be limited through the use V-RG, in order that we might consider some potential future scenarios. First, is it likely that culling of foxes will cease, one of the predicted savings resulting from the successful deployment of V-RG? Farmers are unenthusiastic about local fox populations for more reasons than just the risk of rabies infections of livestock: for example, they are also considered to be a pest to poultry farmers. On the other hand, were the practice of culling to cease, surely the fox population would increase, not only increasing problems for poultry farmers, but also creating higher demands on the efficacy of V-RG. As the epidemiological formula given earlier predicts, an increase in the density of the rural fox population would require a corresponding increase in the percentage of herd immunity to control or eradicate fox rabies.

Another consequence of an increase in the rural fox population if fox rabies were to be controlled by vaccination rather than culling could be their migration to urban areas. Applying the epidemiological formula give above, Professor Roy Anderson (1986) of Imperial College, London has calculated that in urban areas with dense fox populations ($7/km^2$) the required level of immunisation is predicted to be roughly 95 per cent. This means that 95 per cent of the foxes must be immune at any point in time in order to control the spread of rabies. Moreover, in order to maintain herd immunity, given the fecundity and life-span of foxes, vaccination would have to be repeated annually or bi-annually. How could this be achieved with V-RG in an urban setting? Can you *really* imagine helicopters dropping bait (chicken heads were used as bait in one of the field tests) containing a living genetically engineered virus onto European cities? As Anderson (1986) comments, 'These figures are depressingly high and question the practical feasibility of mass vaccination as a means of control' (p. 305). Surely a more likely response to the threat of rabid urban foxes would be to cull them and to vaccinate pets in cities against rabies – all of which would be additional costs incurred as a result of the random vaccination of wild foxes using V-RG. Hence, were a broader, more long-term risk–benefit analysis undertaken, it seems doubtful whether the use of V-RG would actually provide a net financial benefit to the European Union.

Finally, of what significance is fox rabies to human health and welfare? According to Christopher Jackson MEP (1992), 'A rabid fox is only infectious in the three days preceding its death.' Furthermore, unlike the dog rabies of India which cycles between different species, 'the victims of a fox bite can only pass the disease back to another fox. In other words, if a fox bites a dog and gives it rabies that dog *cannot* infect its master with a bite. The disease may only be transmitted to another fox' (Jackson 1992). Human deaths from rabies in the whole of Europe over the past two decades have been of the order of 1–4 per annum, and in France, one of the Member States with a 'rabies problem', there are only 50 dog rabies cases per 10 million dogs and no cases of rabies in humans at all (Wilson 1992). Thus, whilst the threat that fox rabies poses to humans and their pets engenders more fear than it warrants, the use of a genetically engineered vaccine to eradicate fox rabies may engender more risks than fox rabies itself.

MAD DOGS OR JACKASSES?

In July 1974 James Watson – who along with Francis Crick elucidated the double-helix structure for DNA – was one of eleven eminent scientists who asked publicly for a voluntary moratorium on certain types of experiment involving recombinant DNA technology which they considered to be too risky to pursue without further knowledge. However, the moratorium was short-lived, and in 1975 most scientists resumed their genetic manipulation work. When asked in 1979 to explain the apparent *volte face* of these scientists over the risks of recombinant DNA techniques, Watson described the signatories of the moratorium as 'jackasses' (Jahoda 1982).

Would that there were more jackasses around today!

11 Environmental Threats of Transgenic Technology

Dr Sue Mayer

There can be no doubt that transgenic technology fundamentally alters the relationship between humans and nature. Claims from proponents of genetic engineering that this technology does nothing more than speed up traditional breeding methods or evolutionary processes are misleading. No 'traditional' breeding programme could ever result in the introduction of a human or pig growth hormone gene into fish, or insect genes into plants.

It is this basic disruption of species barriers which brings with it the environmental threats of genetic engineering. It is not wrong to say that once genetically engineered plants and animals are released to the environment we will never be quite sure of what is what any more. Once released, living organisms with the capacity to reproduce will not be able to be simply recalled to the laboratory, and the work of evolutionary biologists could become an impossible task.

As the technology is being developed at the moment, the most immediate environmental threats come from the release of genetically engineered plants, microorganisms and fish. The fact that these threats are more immediate does not mean that threats from the release of other genetically engineered organisms, including mammals, do not exist. The same principles will apply to all.

There are legitimate environmental concerns about the impact of patenting genes, microbes, plants and animals, and the continued maintenance of a high-input agricultural system, but these are outside the scope of this review. Here I wish to concentrate on the direct environmental threats of transgenic technology. Having reviewed the environmental threats, I will briefly consider whether the risk-assessment procedure which is enshrined in EC legislation and is the basis of UK legislation will protect the environment, and whether, if carried out properly, large-scale commercial releases of transgenic plants, animals or microorganisms should be allowable.

AREAS OF RISK

There is no doubt that the release of genetically engineered organisms to the environment poses risks. They have been the subject of study by bodies such as the Ecological Society of America (Tiedje *et al.* 1989) and the Royal Commission on Environmental Pollution in the UK (1989).

The risks reflect the uncertain nature of the behaviour of organisms in the environment and the variable interactions between gene expression and the environment. There may, for instance, be the 'escape' of the introduced gene by crossing with wild relatives, unexpected alterations in phenotypic expression in different environments and unexpected advantages conferred by the genetic manipulation which could lead to the establishment and persistence of an organism.

Despite the implication of its name genetic engineering is not a precise art. Integration of foreign (heterologous) DNA into the host genome is a random event. Integration of a heterologous DNA may disrupt other gene sequences and possibly the secondary or tertiary structure of DNA in the cell. It is becoming increasingly clear that the folding and architecture of DNA is important in cell function and that DNA is not in some meaningless jumble. It is easy to envisage how secondary and tertiary structure may influence the binding of regulatory proteins, for instance.

Genetic engineering has thrown up dramatic, visible surprises already. The pigs with an introduced growth hormone gene from cows or humans, produced at the United States Department of Agriculture's (USDA's) facility at Beltsville, showed clinical signs including lameness, lack of coordination, exophthalmus and thickened skin. On postmortem there were signs of gastric ulceration, severe synovitis, degenerative joint disease, pericarditis and endocarditis, cardiomegaly, parakeratosis, nephritis and pneumonia (Pursel *et al.* 1989). Petunias 'engineered' in Germany were found to have an unexpected frequency of a variety of different coloured flowers which has demanded rethinking of our understanding of the effects of inserting several gene copies into cells (see Jorgensen 1990). These are gross changes, easily identified, and therefore unlikely to be commercialised. What is more worrying from an environmental perspective are the much more subtle effects which we will not detect until many years after the release of a genetically manipulated organism (GMO) by which time it will be too late to correct.

Many of the concerns have arisen not only because genetic manipulation has dealt its own surprises but because past experience teaches us that the introduction of exotic species into new environments, either inadvertently or deliberately, has produced many disasters. Zebra mussels in North American waterways, which were introduced via ships' ballast, have displaced indigenous flora and fauna and threaten the survival of some species. In the UK rhododendron is now a pest in parts of north Wales and elsewhere as it has spread from private gardens. This colonisation was not rapid but has gradually progressed, almost unnoticed, until the severity of the problem is such that eradication programmes are both difficult and costly.

The threats that natural systems face from GMOs include genetic pollution, disruption of species composition and diversity and loss of biological diversity.

GENETIC POLLUTION

Genetic pollution of native flora and fauna may come about through cross-breeding between the manipulated organism and a close wild relative. By introducing entirely new genetic material we risk altering the very basis of evolution for these species.

Some crop plants have weedy relatives which may be found close to cultivated plants. Genetic pollution of wild plant species could occur through the transfer of pollen either on wind or by insects. Although pollen transfer rates decline dramatically with progressive distance from the source plant, it has been shown that for insect-pollinated plants, such as radish, gene flow may take place to a weed one kilometre (km) away from a crop plant (Klinger *et al.* 1991).

In field trials with genetically engineered cotton, another insect-pollinated (largely bumble bee) species showed evidence of gene flow to surrounding commercial cotton at low levels (less than 1 per cent) at distances of up to 25 metres (m) (the furthest distance investigated) from the transgenic cotton (Umbeck *et al.* 1991). In potatoes, distance of pollen spread from engineered to neighbouring plants was much smaller, being restricted to 4.5 m (Tynan *et al.* 1990).

The frequency of such an event may be small, and often the foreign gene will not become established, but it is extremely unlikely that we will be able to predict with sufficient accuracy those situations where the risk is greatest. And with the field release of GMOs on a commercial scale, rare events may well occur. In all these cases the prevailing environmental conditions and uncontrollable factors, such as abundance of insect pollinators, will critically affect the outcome.

Although it seems superficially less likely that such a situation would occur with the majority of animal species currently being engineered because domestic livestock are relatively easily contained and experimental animals are mostly housed in secure sites, there may not be 100 per cent containment. It is easy to envisage the escape of a bull, cow, ram or ewe which, even if only a short time elapses between escape and recapture, mates with a neighbouring farmer's stock. Whilst this may not be a real environmental threat in its strictest sense, it does pose the spectre of high-security farming with special fencing stretching across the countryside in an attempt to keep the genetically engineered genes and animals in. And such precautions would be taken in the hope that someone does not leave the gate open, or that sheep do not continue to demonstrate their abilities to get through fencing.

There is, however, an example of a practice in the area of genetic engineering of animals where the genetic pollution of a natural species is inevitable if it is allowed to continue, and that concerns genetically engineered fish. Species such as trout and salmon are being genetically engineered to grow more rapidly by inserting growth hormone genes from other species, or to be more resistant to cold by the transfer of genes from species of fish which have evolved to live in colder seas. The genetically engineered fish are destined for fish farms which often comprise floating cages located in sea lochs or along the coast. It is well known that fish escape frequently from such systems, and it is thought that our 'natural' populations of salmon and trout already will have been affected by the influx of farmed fish. For instance, in Norway disease has spread from farmed to wild fish. An influx of foreign genes can hardly be welcome and will make the nature of a fisherman's catch much more uncertain.

DISRUPTION OF ECOSYSTEMS

There are a number of ways in which ecosystems could be disrupted as a result of genetic engineering practices. For example, the 'escape' of an introduced gene to a native species may result in the appearance of a new pest by giving the native species an added competitive advantage over other organisms and result in their being displaced. The transfer of a gene encoding herbicide-resistance to a weed species could render it resistant to that herbicide, and that could result in a new herbicide being needed to control weeds. In fish populations, the transfer of growth hormone gene could result in the emergence of a strain of fish which is much larger and which outcompetes the 'natural' strain of fish for food.

There may be disturbance because some ecological balance is upset by the mere presence of the engineered organisms. An advantage conferred by the foreign gene may allow an organism to persist. The way oil seed rape has spread into the hedgerows since its widespread cultivation began some 10 years ago indicates that this is possible and, as the Royal Commission on Environmental Pollution (RCEP 1989) noted, it is extremely difficult to predict why one species becomes a pest when a close relative is not, and why an organism may be rare in one ecosystem but a major pest in another. For instance, *Vulpia* is a rare native grass in the UK but a major introduced weed in Australia.

In the case of fish, damage could result even when so-called environmental protection mechanisms are used. For example, efforts are being made to engineer male fish to be sterile. However, were they to escape into the environment and were they in someway bigger or better able to compete for food (because of the effects of an exogenous growth hormone gene for example), then they may be the fish most likely to gain access to the females' eggs for the purpose of 'fertilisation'. Given that the engineered fish are sterile, the population of native fish could be seriously depleted.

Once microorganisms are released they too may become established. We know less about microbial ecology than we do about plant ecology, and although 'suicide' mechanisms are being built into many microorganisms the likelihood of reversion mutation remains unacceptably high. Microorganisms will not be released in their ones and twos but in their millions. The selection advantage of a mutation which reversed the 'suicide' mechanism would be great.

Whether such 'suicidal' or 'disabled' organisms will be able to perform the function for which they have been engineered seems unclear, and in some cases such organisms have not been sufficiently persistent in the field trials. On the other hand, the dangers of introducing microorganisms which are designed to persist are likely to be great.

There is also the possibility that when viruses are engineered to have broader host ranges they may not only damage the pest they are designed to target but also affect beneficial organisms.

LOSS OF BIOLOGICAL DIVERSITY

The effect of genetic engineering on biological and genetic diversity has provoked much controversy. On the one hand, proponents of the technology claim that more genes will be available since the genetic material of the entire kingdoms of organisms can be mixed and matched. On the other hand, opponents claim that identical genes will be spread across different species and that only existing crop varieties, which already have a very high degree of genetic uniformity, are being engineered and thus little of the claimed 'mixing' of the gene pool will actually occur.

Biological diversity and genetic diversity are not, in fact, the same thing. Genetic diversity has been defined as the distinct variation within species (Reid 1992). Biological diversity, however, includes the variation that occurs between habitats and can be equated with species richness, that is the number of species, plus the richness of activity each species undergoes during its existence through events in the life of its members, plus the non-phenotypic expression of its genome.

It is already quite clear that the applications of genetic engineering could not be claimed to increase diversity in either genetic or biological terms. In fact, the technology of genetic engineering cuts across all the natural evolutionary processes which have formed the basis of biodiversity. Trying to 'make' diversity is likely not only to result in the continued loss of diversity as crop and animal varieties are made more uniform, but such an approach does not tackle the basic problems of loss of more diverse local crops and breeds of animals, or the continuing loss of habitat.

In fact genetic engineering can only result in a system which is, and can never be other than, the very antithesis of a biologically diverse system as it attempts to impose a specific, human-chosen genetic composition on natural systems.

PROTECTING THE ENVIRONMENT

In contrast to older technologies, such as nuclear energy, the damage that could be caused by genetic engineering is being predicted rather than observed after the event. Legislation is being introduced to control the environmental damage the technology may pose rather than attempting to retrieve the situation. This poses a very real test of the precautionary approach to environmental protection. Although there is no categorical proof that release of a GMO will be damaging, there is no proof that it is safe. For once, we are trying to prevent damage being done.

Will the step-by-step, small-scale field trial approach to regulation result in the production of sufficient scientific information to allow all the questions which are considered necessary to be answered before a commercial, full-scale release is undertaken? From a strictly scientific basis, the answer cannot be yes. There is huge uncertainty and ignorance in our knowledge of the difference in behaviour of organisms in field trials compared to the wider environment. It is difficult to envisage how licences given in the UK will ensure the safety of the organism under all the widely differing environmental conditions across Europe. [Under the environmental release directive (90/220/EEC) a consent to place a GMO on the market will apply throughout the European Union (EU).]

Scientists have already identified this problem. Umbeck *et al.* (1991) concluded from their studies of pollen spread from genetically engineered cotton that there is probably no economically feasible or totally ecologically safe field design that will provide natural growing conditions and contain pollen completely.

Not only, therefore, are current field trials not fully contained, but natural conditions are not being met. In relation to microorganisms, Morgan (1990) has stated that safety may have to be determined directly in the environment, since data from the laboratory or enclosed systems may not be suitable for predicting the behaviour of GMOs after their release. Multistage tests aim to limit the uncertainties in order to minimise the risk. However, a position where there is no risk cannot be achieved.

From the above it can be inferred that committees advising on the safety of releases and the industries that are seeking to commercialise GMOs will not be able to call on 'hard' scientific evidence to justify their decisions. They will be making political decisions about what uncertainty is acceptable in their judgement of the risk. They will have to decide that our level of ignorance is small enough to have made the decision safe, and that they are confident that the inevitable unquantifiable risks arising from mistakes or irresponsible use are worth taking, and that the benefits really do outweigh the risks of disturbing the integrity of nature.

These will be difficult decisions, especially in a climate of public opinion which wishes to see the environment protected and nature no longer defiled. Legislation is being implemented in the UK with unnecessary secrecy, a lack of adequate representation from public interest groups, no requirement for consultation, and influence being given to those who have most to gain – the biotechnology industry. This, coupled with the apparently tight remit of the Advisory Committee on Releases to the Environment (ACRE) – only to consider the effects of the release in question, out of context of its wider impact – does not bode well for environmental protection.

Greenpeace believes we do not need to take the risks engendered by the release of GMOs to the environment. GMOs are being promoted as 'technical fixes' to environmental problems which have been created by exactly the same attitude that has spawned genetic engineering. We need to be developing alternative solutions which do not seek to control nature and threaten evolution but which address the fundamental problems. The biotechnology industry has lost touch with how people want our new environmentally aware society to evolve, and despite their claims they have no 'hard' science to hide behind.

Discussion II

Phil Brook: Is an animal being kept for spare-part surgery likely to be kept in an antiseptically sterile environment, and if so, is the environment likely to be a behaviourally sterile one?

Prof. John Webster, University of Bristol: Not necessarily. The animal must be immunologically compatible with a human but it need not, necessarily, be completely germ free. The complex immunology of this is beyond me and many others at present, but histo-compatibility with humans is not necessarily incompatible with immunological resistance to natural pathogens in the external environment.

Phil Brook: On the other hand, presumably you dampen down the animal's immune system?

Prof. John Webster, University of Bristol: We are not making these animals immunologically suppressed. Rather we are trying to make them immunologically compatible, which is not the same thing as immunosuppression.

Helen Nelson: Professor Polge referred to transferring the embryos from cows. He said that they were flushed out and that no surgical procedure was needed and that it was quite painless and simple. So I wonder if someone could enlarge on that please?

Prof. John Webster: In cattle, for the most part, the collection and insemination of embryos can be achieved through the cervix. With sheep such processes involve laparoscopy. There is a degree, obviously, of interference with the animal in both cases, which one can assess. In sheep there is the equivalent to abdominal surgery, the laparoscopy, in the human. In cattle, going through the cervix is sometimes said to be the same as artificial insemination. I said it is almost the same but it is not because neither procedure is undertaken at the time of oestrus when the cow is on heat and the cervix is open. In fact epidural anaesthesia is used for these procedures, which implies that there has to be some degree of discomfort. I am being very precise in what I am saying. I am trying to deal with all questions rationally, avoiding paranoia at all costs. There is some degree of mild distress to some dairy cows caused by this process. There is rather more moderate distress caused to sheep.

Anon: What is the surgical procedure?

Prof. John Webster: Incision with a knife. In the case of a sheep it involves two stab incisions in the abdomen.

Helen Nelson: I feel that we have been misled by the scientists because it *does* involve a certain amount of distress to the animal. The epidural is also a surgical procedure on a slightly smaller scale, and so in the cases of both cows and sheep it is a form of surgical procedure.

Prof. John Webster: When these procedures were done for advancing scientific knowledge they had to be done under the Animals (Scientific Procedures) Act 1986. Now they are routine with cattle they are, in effect, done under the new version of the Veterinary Surgeons Act 1966. The point I am making is that at the moment farm animals do not enjoy the same protection in law as laboratory animals. Laboratory animals do not have that much protection, but farm animals have even less.

Anon: Could I make just a general point of information following on a comment by Professor Webster that farm animals do not yet have the degree of protection provided to laboratory animals under the Animals (Scientific Procedures) Act 1986. I would just like to make the point that this Act provides a great deal of protection to laboratory animals on paper, but does not provide any actual protection to a laboratory animal in a real laboratory because: (a) the legislation only requires a closed, not publicly accountable, committee formally to endorse activities by its own members; and (b) what procedures do exist for disciplining failure to follow regulations are simply not enforced, as was demonstrated. I take as my example the case at the Medical Research Council's (MRC's) facility at Mill Hill, where outsiders were able to demonstrate that animal welfare procedures required under the Act were simply being ignored by the people who had the responsibility for enforcing them. So I think it would be very dangerous to say we ought to bring farm animal protection up to the same level as laboratory animal protection as that would only mean a purely paper public relations level of protection.

Anon: May I make just one very simple point. The fact that people break a law does not necessarily make it a bad law.

A bad law is one where it can be shown that it cannot be enforced for the simple reason that all the law requires is the endorsement of procedures. Whilst the Animals (Scientific Procedures) Act 1986 was going through Parliament, a former official of the Research Defence Society [a pressure group in support of the use of animals in scientific research] assured his membership that not a single animal experiment would be prohibited under the legislation. So we have the assurance from the main pressure group defending the interests of the animal experimentation industry that the legislation would have no effect on their activities, and I think he has been proved right. After all, the number of experimental procedures involving the use of animals has increased over the last year.

Dr Alan Long, Vegan Research: I would suggest that we do not have to use animal systems. I would like to ask Professor Burke when does he think it will be possible to avoid using transgenic animals and use transgenic bacteria instead?

Prof. Derek Burke, Advisory Committee on Novel Foods and Processes (ACNFP): A number of products can be produced in both bacterial cells and eukaryotic cells in culture. However, there are two sorts of problems associated with this. One is improper folding of the resultant protein with the result that very often you get a very low yield of soluble products.

The second and more important problem is that certain proteins must be modified, for example, by the addition of carbohydrates, and such modifications are essential if the product is to be biologically active in human cells. So, whilst you can make human Factor IX in bacteria and you can make it in eukaryotic cells in culture, without knowing the economics my guess is that the cost is high and the supply is limited. On the other hand, costs will fall if one uses a lactating system like the one I described. So the issue before society is a cost–benefit analysis. Do the hazards, if you like, of using an animal for this process, which I have tried to identify and describe, outweigh the reduction in cost and, therefore, the greater availability of human Factor IX not only for a very affluent population but for a population of middling economic wealth?

Florianne Koechlin, European Co-ordination, 'No Patents on Life', Switzerland: I would like to raise a risk arising from transgenic animals. If you put a human gene into a swine, for example, you create a kind of evolutionary opportunity for microorganisms and viruses causing swine diseases to mutate. In the worst case it is possible you may create human pathogens. We must bear in mind that viruses from different species are very similar to each other, for example the immunodeficiency virus of apes (SIV) and the immunodeficiency virus of humans (HIV) are very similar. We must also bear in mind that it sometimes takes 10 to 15 years for such diseases to break out.

Dr Sue Mayer, Greenpeace UK: I agree that there are types of risks which are theoretically very real but upon which we have no way of putting a quantity because we do not know anything about their probability. We have no information and so what is happening at the moment is that we are just guessing the risk, and we have not had the discussions to decide whether we want to take that risk.

Prof. John Webster: I think that is an overstatement personally, and we will get differences of opinion from people on the platform. But it is worth pointing out that the growth hormones of the humans and the transgenic animals concerned are almost identical in structure, they are non-immunogenic, and so the human gene product is very similar in its structure and its general effects on the transgenic animals.

Dr Donald Bruce, Society, Religion and Technology Project, Church of Scotland: My own background is actually in nuclear safety and it is intriguing to see the parallels between the genetic engineering issues we have been discussing today and the nuclear energy issues. One of the things I have learned in risk assessment in nuclear engineering is that there is no such thing as zero risk in anything. There is no such thing as a totally safe procedure. It does not exist. But neither is it possible to claim that there exists a method of risk assessment which can quantify all risks. However, just because you cannot quantify a risk, does not mean to say you cannot assess it. In my opinion, one should not get too 'hung up' with numbers. We spend all our lives assessing risks in non-quantitative ways; moreover the numbers may not be very well calculated in the first place! Finally, one should not let one's risk assessment be dominated by highly improbable, but very high consequence events, because if the assessment is pushed too far on that basis, one will skew the whole assessment.

Anon: I think you are right when you say you can use the knowledge that you have already gained to make an informed assessment of risk. But when you make an assessment of any risk you also have to be open to the fact that you may be ignorant of certain areas, and what should be explicit in that assessment is what we do not know. Risk assessment should not be presented as a scientific activity when it is not.

The second point is that risk assessment should not be something that is done in private by a committee which has no public consultation procedure, no allowance for the public to input into the decision-making process. There should be adequate information for the public to participate in decisions involving risk assessment.

Albrecht Muller, Germany: I want to add a contribution to the discussion of the risk assessment of rabies vaccination. As a result of the rabies vaccination programmes, the fox population has got more dense as a consequence of which we have seen more cases of bentworm, which is very infectious to people and very often ends with death. I raise this point as an illustration of how difficult risk assessment is.

Ruth McNally, Bio-Information (International) Limited, London: I agree, although an increase in the fox population should have been predicted. If you vaccinate foxes as a way of controlling the spread of rabies instead of culling them, then the fox population will increase and so will the incidence of pathogens for which the fox is a host reservoir. As you observe, it is unlikely that all of these pathogens will have been taken into account in the risk assessment. Moreover, unless the level of herd immunity in the (expanding) fox population is maintained by regular, probably annual, mass vaccination programmes, the level of the fox rabies virus will itself increase with the size of the fox population.

Dr Jan Staman, Ministry of Agriculture, Fisheries and Nature Mangement, The Netherlands: It seems to me that we are talking today about three types of risk arising from the genetic manipulation of animals: animal health; animal welfare; and pollution of the environment. That is what concerns us. My question is: Is that what concerns the general public? Is it not that the public is worried that scientists are next to God? Were we to remove the risks to animal health and welfare and environmental pollution, would the British people then be happy? That is my question.

Dr Tim Lang, Parents for Safe Food: I am interested in current experiments in Denmark, done by and under the auspices of the Danish government, on how to communalise decision-making and risk assessment. There they selected a representative panel of Danish people, comprised of 12–20 people, before whom a spectrum of experts on a particular technology was brought to present cases 'for', 'against' and 'in between'. Then the panel had to write a report which it had to justify to the same experts and representatives of different positions who had presented their cases to the panel. The outcome of the 'consensus conference' was fed into the Parliamentary procedure.*

With an enormous technological revolution, such as biotechnology and genetic engineering, we must experiment with our processes for making decisions and judging risk assessment, and for involving the public. Otherwise risk assessment somehow sits in judgement as a neutral technology over the public and over the process.

Anyone who has looked at risk assessment knows that it is not a neutral technology. One must breathe life into it. Whether we come from industry or government, as academics, consumers or environmentalists, we must work out how to

* The first UK National Consensus Conference took place in November 1994 on plant biotechnology organised by the Science Museum, London.

breathe life into these the processes of risk assessment, which currently, and particularly in Britain, are fossilised.

Prof. Derek Burke: I agree with much of what Tim Lang has said, but he is, I think, overstating the fossilisation of the British position. He is also understating the need for technical evaluation of some issues. I think you would agree that in risk assessment the issues range from the purely technical, which only a working scientist can probably formulate, right through to social issues.

I am in favour of experimentation, I hope my paper made that clear. But my advice to those of a different persuasion is, do not set the committees up as 'Aunt Sallies', because there are people on the committees who are not unsympathetic to what you are trying to do. Learn to work with the system we have got, because we cannot throw it away. Promote evolution rather than revolution would be my advice.

Dr Jan Staman: Whilst campaigners would like to have a moratorium, there is the opinion in the Netherlands that we would thereby miss quite a lot of experience in the genetic modification of animals. I do not mean the technological experience, I mean the ethical experience. If you have a moratorium then you are in the land of nowhere – nothing is happening. So, now my question is for you, Dr Lang. What would be, for you, the proper moment to stop the moratorium? When other countries have articulated the ethical problems? Or is your moratorium just the beginning of a stop forever? What for you, Dr Lang, would be the point to end a moratorium and to begin normal science and normal production?

Dr Tim Lang: I hoped someone would ask that. We must start looking for, if not middle ways, ways between the extremes. Not because one does not have the right to be at either extreme – on the contrary, I think that that is a very important right – but because, inevitably, unless the consumer movement (and I am speaking for the consumer movement here) starts getting into that middle ground it will be sucked into the extreme of being permissive to industry and that, I think, in the long-term, means the loss of any independence of the consumer movement. We have had 15–20 years of permissiveness and, belatedly, some veneer of consultation – which I welcome! But the main concern for consumers is the avalanche of new products. Somewhere along the line, industry and regulators must accept the moral, political, and economic responsibility for restraining that avalanche. Now, you note I did not say a ban. What I said was a moratorium on development in order to allow time for public education.

Anon: Several of the speakers have brought up the issue of committees. These committees are now forming fast and furious, but on not one of them do I see representation of the people who are most affected by the whole technology, namely the farmers. The farmer with mud on his boots, the actual stockman who handles animals and understands about them, does not sit on any ethical committee, biotechnology committee or any other committee that will be assessing the ethics or advisability of the genetic manipulation of farm animals, and I think this needs to be righted.

Dr Kevin Ward, CSIRO: I think this is an important point and one which has been addressed in Australia. We do have farmers on our ethical committees and on what we call GEMEC, which in Australia is the authority which controls this work.

Anon: A point of information. The UK Advisory Committee on Releases to the Environment (ACRE) does have a representative from working farmers.

Joyce D'Silva: When the press release from the Ministry of Agriculture, Fisheries and Food (MAFF) arrived on my desk saying that there was going to be a study group on the ethics of genetically engineered food, I thought, 'Gosh, things are moving'. However, when I talked to Professor Burke today, he told me that the group is only concerned with the ethics of feeding genetically engineered foods to people, and is not concerned with what they do to animals or anything else to make these foods. So Mr Gummer's [UK Government Minister for Agriculture, Fisheries and Food] stated commitment to animal welfare looks rather shoddy.

Prof. Derek Burke: But at least you are getting a study group looking at an ethical issue driven by some science, from which I thought you might derive some sense of satisfaction.

Joyce D'Silva: With all due respect, I find it unfortunate that the press release issued by the MAFF does not give the study group's true terms of reference as you describe them today.

Anon: I would like to ask Professor Burke how he can honestly recommend to a conference such as this, the work of the Ministry of Agriculture, Fisheries and Food's (MAFF's) ethics study group whose membership and terms of reference are so obviously intended to exclude consideration of the ethical implications for animal welfare of genetically modified food.

Prof. Derek Burke: You misunderstand the situation. I am asked by Government to do a job with particular terms of reference. Now I take on that job because I am concerned about the social issues we are talking about; that is why I am here this afternoon. And I am trying to work at the system and I think we have made some changes.

The questions you want to have addressed are on a broader political platform. I do not have the power of a senior politician. All I am saying to you is that the system is changing. Do not set up the scientists on committees and study groups as the 'fall guys', because they are not unsympathetic to some of the things you say.

Anon: Thank you for that, but I do not think you actually answered the question. The questioner was concerned that the MAFF's ethics study group was not set up to address issues of the ethics and welfare of genetically modified farm animals.

Prof. Derek Burke: I was not consulted about the terms of reference of the job. Those are political decisions taken by a Minister. I do not have the authority of the Ministry of Agriculture, Fisheries and Food. It is a campaign issue.

Anon: I am baffled by Professor Burke who says that the terms of reference of the various regulatory committees are inappropriate and inadequate, and yet he wants us to lend legitimacy to them. And he asks us not to set up the scientists on such committees as 'Aunt Sallies'.

Prof. Derek Burke: I did not say that.

Anon: That is the implication of what you said.

Prof. Derek Burke: No. That is not the implication of what I said. You think they are inappropriate. But I am saying I have been invited to work within the terms of reference of a particular committee. The ethics study group referred to arose out of a specific problem that our committee – the Advisory Committee on Novel Foods and Processes (ACNFP) – identified, and I thought we might have got a little credit (but apparently not) for actually 'blowing the whistle' on what we felt to be an ethical issue and not shoring it up behind closed doors. We have really tried to open this issue up: we have written to a number of the people who are represented here, and I have personally made sure all the alternative groups were consulted. Now it is not the consultation you want, because it is not broad enough, but it is at least a consultation. That is all I will say.

Richard Ryder, RSPCA: Just a point of information. How many committees already exist, and how many committees may there be in the future, to address this issue?

Dr Tim Lang: In the UK we have: the Health and Safety Executive's Advisory Committee on Genetic Manipulation (ACGM); the Department of the Environment's Advisory Committee on Releases to the Environment (ACRE); the MAFF's ACNFP of which Professor Burke is the Chair; the new MAFF ethics study group; the Food Advisory Committee (FAC); the Farm Animal Welfare Committee (FAWC); the Department of Health's Committee on Medical Aspects of Food Policy; the Committee on Toxicology; and the Home Office's Animal Procedures Committee (APC).

Anon: Is this not part of the problem, Mr Chairman? We have actually got a proliferation of committees which is bound to lead to confusion. These committees ought to get together and have a conference of their own.

Prof. Derek Burke: Just as a point of information, we do have to have joint meetings Mr Chairman.

Clare Norman, Guildford, Compassion in World Farming (CIWF): We must not forget the animals. There was a quotation in one of the *Agscene* magazines that said, 'The question is not can they talk, nor can they reason, but can they suffer?'

Dr Donald Bruce: I would like to suggest that science should be socially responsible, and that therefore it has ethical dimensions.

Prof. John Webster: Yes, it is wrong to suppose that science is amoral. In practice, the scientist has always got an ethical assumption of some sort about what he is doing. I would add that the morality of scientists – the individuals who perform science – is very variable. Moreover, the disinterested pursuit of knowledge as an ideal to which we all strive but never achieve is, in fact, an amoral pursuit.

Dr Donald Bruce: I want to say, as a Christian, that the idea of reducing the cognition of a species is reducing something God-given.

Prof. John Webster: That terrifies me too. As I said, I realise that applying strictly scientific criteria to the ethical limits of genetic engineering gets my own mind crashing up against the buffers too.

Dr Alan Long: The title of our conference is: 'Animal Genetic Engineering: Of Pigs, Oncomice and Men' (which I take to mean both men and women). Now, we have heard about pigs and oncomice and I am wondering why we always expect to adapt animals to our convenience? Why not modify ourselves a little bit so that we are better inhabitants of this world and put a lighter footprint on the planet?

I have asked myself, 'If I were to have some genetic engineering done to me, what would I have done?' And the answer is that I would quite like to tweak up my genome and become a better herbivore – a better cellulose converter.

I would like to ask the team what they would like to have modified in their genes?

Prof. Derek Burke: A long time ago Faustus made a choice not to age. I suppose many of us would make the same choice, but it is not technically feasible. On your choice, Alan Long, for several years there have been experiments at Newcastle University on making transgenic pigs that might be able to eat grass, so, who knows?

Dr Alan Long: So in the future I could happily eat grass mowed from my own lawn!

PART III

Patenting of Genetically Engineered Animals

12 Patenting of Transgenic Animals: An Industry View

Meredith Lloyd-Evans

The industry view on the patenting of transgenic animals can be expressed simply: patent law, as it stands, does not prevent in principle the patenting of an innovative and useful product of human discovery and ingenuity. From this point of view, a transgenic animal, or at least the gene construct within it, is eligible for patenting. The basic principles of patenting law have not so far been challenged in court, although decisions by individual patent offices have been.

Moreover, industry in all successful systems is based upon capitalist principles. Investors, whoever they may be, expect a return on investment and expect a company to prosper in the market by exploiting positions based on closely guarded intellectual property or creative marketing techniques or both. From this viewpoint, industry is convinced that without the monopoly position implicit in the granting of a patent, the heavy investment required to discover, develop and make available a transgenic animal and its products cannot be justified.

Activist groups opposed to this have several realistic avenues to explore. One is to construct with industry a feasible alternative to the capitalist system, where success is not dependent on a monopolistic market position. Another is to contribute to an ethical position, which is expressed within national or supranational laws, that prevents transgenic animals from being developed or, if patented, being marketed. A third way is to seek to change patent law, an avenue being pre-empted by recent ethics-orientated decisions by the European Patent Office (EPO) and Japanese Patent Office (JPO).

From the viewpoint of biotechnologists, there are several benefits to humankind which accrue from the development and application of transgenic animals. There may also be benefits to individual animals or animals in general. Industrial scientists regard it as inconceivable that activist groups should prevent such positive developments.

The 'industry' could be said to consist of farmers, companies responsible for farming inputs such as animals and animal medicines, food processors and finished food providers. From industry's viewpoint, patent protection is an essential incentive for biotechnology companies. The guarantee of market protection is vital in order to earn the profits necessary to fund essential research and development, and the lack of protection for intellectual property will result in humanity missing out on potentially life-saving innovations and solutions to many global problems.

I am concerned with only those input suppliers involved in biotechnology. My viewpoint is that a constructive dialogue is possible between industry and activist groups and that both parties should work harder to resolve the existing destructive conflict. Lines may indeed need to be drawn somewhere, but not at extremes.

PATENTING – THE LEGAL FRAMEWORK

The legal framework for patents is relatively simple and with some variation is the same all over the world. An invention is patentable if it: is truly novel; is the product of human ingenuity or discovery; has industrial utility; is described in such a way that the 'person skilled in the art' can make it work; and is not 'obvious' from previously published work or general knowledge.

Patenting is a system of invention registration which gives the inventor a limited period of time during which he or she can defend the patented idea or product through courts of law against other people or companies who steal the idea and sell infringing products. In return, the inventor reveals the details of the innovation to the world at large through the publication by the Patent Office of the patent document. Patenting is carried out not only by large or multinational organisations. Many significant inventions that have led to progress in society have been the work of individuals (Edison, for example), and the costs of establishing a useful patent are not astronomical.

It is worth stating that ideas and inventions do not have to be patented. They can be revealed without restriction, as often happens with advances made by individual scientists. They can be kept as confidential knowledge by companies, which is often the best way to keep processes and engineering advances under company control. Even when patents have been established, patent-holders can choose not to charge royalties or can be prevented by government orders from charging royalties on the licensing of their patents.

PATENTING – THE ETHICAL FRAMEWORK

There has always been some ethical consideration in the granting of a patent. Generally speaking, new methods of causing public harm and disorder (such as a method for mass murder or a method of encouraging sexually immoral acts) are prohibited. With the advent of innovations involving animal life, there are wider ethical implications. The European Patent Convention (EPC) bans patents on inventions contrary to public order and on immoral inventions. In making a decision on the patentability of an invention from an ethical perspective, the EPO applies ethical risk benefit criteria: is the potential benefit of the invention to human health outweighed by the likely animal suffering?

The decisions by the EPO in regard to the 'oncomouse' (positive) and a mouse developed to test hair-growth agents (negative) illustrate the increasing attention paid to this area. In both of these cases, informal risk or harm benefit analyses were used. It may only be a matter of time before an applicant is requested to provide a more formal ethical risk benefit analysis as part of the support for the patent application.

PATENTS ON LIFE – THE CURRENT SITUATION

Patents on life are not new, though they are relatively modern. Since the granting of the first patent on life in 1873 by the Patent Office in the USA (US PTO) to Louis Pasteur for a purified strain of yeast, tens, if not hundreds, of thousands of microorganisms and their uses have been patented. The first patent for a living vaccine agent, another substantial field of patents on life, was granted by the US PTO in 1904. Through antibiotics and vaccines these patented life-forms have saved millions of lives, and have fed hundreds of millions through yoghurts, bread and other fermentation foods and drinks.

Gene constructs and genetically manipulated microorganisms (GMMOs) and cells are also patented, for example gene sequences from whooping cough bacteria, which might be used to produce a diagnostic agent or vaccine, gene sequences related to house-mite allergens, and gene sequences for the production of wound healing factors. Currently it is possible to patent microorganisms producing novel compounds, gene constructs, and GMMOs. However, it is not possible to patent plant varieties which are protected by the International Union for the Protection of New Varieties of Plants (1961) (UPOV).

UPOV was established in 1961, at a time when the potential application of biotechnology to plant life could not have been imagined. Since the advent of recombinant DNA technology in 1973, patent protection first for gene sequences and techniques used in plants, and then for the transgenic plants themselves, has increasingly been sought through the patent system. A recent amendment to UPOV allows a plant to be both UPOV-protected for its characteristics and patent-protected for the technology used to produce these. Recent examples of patented plants include crops that have enhanced amino acid content, tomatoes with anti-insect genes and trees that resist serious virus infections.*

Patent law regards animals as part of a continuum of living organisms, beginning with micro-animals. There are no explicit bans on the patenting of animals in over 50 countries, and patents have been granted in the USA and in Europe for transgenic animals. Patenting of animals is permitted in Australia, Argentina, Canada, Greece, Hungary, Japan, the Netherlands, New Zealand, Switzerland and Turkey. Patenting of transgenic animals is permitted in the USA and EPC countries (subject to the provisions of the 'ethics' Article 53 (a) of the EPC) but is banned in Denmark. Some countries have clear ethical positions with regard to transgenic animals, and this provides a leverage point for discussions in other countries on whether legal bans will realistically filter out non-beneficial developments.

---

* Editors' note: The EPO grants patents on plants and animals, provided they are not 'varieties'.

HISTORY OF THE PATENTING OF GENETICALLY MANIPULATED ORGANISMS

The number of applications for transgenic animals and associated techniques remains minuscule compared with those for genetically manipulated microorganisms (GMMOs) and associated techniques: of 4116 applications in the general field of biotechnology to the EPO in the 4 years to the end of 1989, less than 1000 concerned genetic manipulation and only nine patent applications for transgenic animals or techniques associated with them had been received. Although there are rather more patent applications on file in the USA, the world is not being overwhelmed with patent applications for transgenic animals at the moment. A brief chronology of events relating to the patenting of genetically manipulated organisms is given in Table 12.1.

The decision-making associated with Harvard University's patent application for 'oncomouse' is of some interest. The University team had transferred into a well-known laboratory strain of mouse a gene (an 'oncogene') which in humans was known to be involved in the onset of breast cancer. The mice proved to develop human-type breast cancer and therefore appeared to be the answer to current great difficulties in assessing possible human breast cancer treatments. In other words, the mouse (known as the 'oncomouse') is a genetically engineered model for a human disease.

In its patent application to the EPO the University made claims for the method, broadened to include genes promoting any type of cancers in any type of non-human mammal, and claims for the genetically engineered animals themselves. The EPO Examining Division rejected the application in 1989 on the grounds that the patent was too broad, and that the EPC did not permit the patenting of animals. In 1990, the Technical Board of Appeal of the EPO disagreed with the decision of the Examining Division, and stated quite firmly that the EPC allowed animals to be patented provided they were not animal 'varieties' (for example, Large White pigs, or Lop-Ear rabbits), and that one example was sufficient to support broad claims if there was no evidence to suggest wide relevance could not be expected. However, the morality of granting the patent was questioned.

Table 12.1
A Chronology of the Patenting of Genetically Manipulated Organisms

1973	European Patent Convention (EPC) states that European patents shall not be granted for plant or animal varieties but can be granted for microbiological processes or the products thereof.
1980	USA Patent and Trademark Office (US PTO) grants first patent to genetically engineered bacteria.
1984	First patent application for a genetically manipulated animal (GMA) lodged with US PTO – Harvard University's 'oncomouse'.
1985	US PTO decides that plants are patentable.
1987	US PTO grants first patent to a transgenic animal, an oyster with increased chromosome copy numbers leading to quicker growth.
1988	US PTO grants its first patent on a transgenic vertebrate to Harvard University for the 'oncomouse'.
	European Commission's proposal for a Directive on the legal protection of biotechnological inventions published.
1989	Bill proposing a 2-year ban on patenting transgenic animals defeated in US Congress.
	European Patent Office (EPO) Examining Division rejects 'oncomouse' patent application as inadmissible under the EPC.
1990	EPO Technical Board of Appeal decides that EPO Examining Division should reconsider its verdict on the 'oncomouse' patent application.
1992	EPO Examining Division grants patent on the 'oncomouse'.
Editors' Update:	
1993	16 legal Oppositions to 'oncomouse' patent filed at EPO.
1995	European Parliament rejects proposed EU Directive. Oral proceedings to 'oncomouse' patent to be heard in Munich by EPO's Opposition Division.

The EPO Examining Division reconsidered its decision and granted the application in 1992, stating that the benefit to humans was likely to outweigh the possible suffering to the mice, and noting that it was not appropriate for patent law to be used to decide questions of morality [see also Chapter 13].*

BACKGROUND TO INDUSTRY'S VIEWS ON PATENTING ANIMALS

There are a number of opportunities that researchers and companies see as justifying the effort and funding put into the development of transgenic animals.

Genetically manipulated animals (GMAs) could be used for human healthcare, as models for disease (e.g. the Harvard 'oncomouse') or as producers of therapeutic substances (e.g. pharmaceutical proteins from transgenic 'pharm animals'). In agriculture, GMAs could be developed: which resist disease better; which make better use of nutrients, reducing waste, helping return more foods to human use and reducing the environmental impact of farming; and which produce more food, milk, eggs, wool for human use. Companion animals could be developed which are disease resistant.

Overall, developers and exploiters of gene techniques believe these advances will reduce animal suffering, enhance

---

* Editors' update: By February 1993, The EPO had received 17 formal legal Oppositions to the 'oncomouse' patent, including one from CIWF and BUAV supported by 35–40 animal welfare societies. (The European Parliament also passed a Resolution calling on the EPO to revoke the patent.) Oral proceedings, scheduled for November 1995, should result in a ruling by the EPO's Opposition Division in favour or against the Oppositions. This ruling could itself be the subject of further appeals.

A further development was the EPO's decision in February 1995 to uphold part of Greenpeace's objection to a patent for herbicide-resistant rapeseed. The objection challenges the EPO's interpretation of 'variety' and the distinction between a 'biological' (unpatentable) and 'microbiological' (patentable) process. This decision and the rejection of the EU proposal for a Patenting Directive (see chapter 13) may result in pressure to redraft the EPC.

human health, and benefit the environment far more than existing and conventional technologies might.

Industry has the view that patents are as essential in these areas as in others to assure the developers that they can recoup some, or all, of the development expense by exploiting a favourable market position.

There are other possibilities that are not, in my view, acceptable and should be avoided by cooperation and agreement between researchers, industry and interest groups. These include: farm animals that have lost the anatomical characteristics of their species such as wings, or have gained new and abnormal features such as a second udder; and companion animals which have been engineered into new colours or forms (for example as suggested by some Southeast Asian sources in the context of decorative fish breeding).

One issue for both industry and activist groups is where to draw the line. We should be identifying developments in GMAs, however few and limited, which the majority of interest groups believe are supportable, and others which all parties agree are not defensible and should not be explored. Industry is willing to work on ethics frameworks that foster beneficial developments and suppress those that damage animal or human welfare. It does not want to be judged on any past record, especially as the whole climate of social responsibility has changed. It is time for new approaches.

There is another issue that will not be answered by a ban on patenting. Many groups attack developments in animal technology, whether bio-technology or otherwise, because they have fundamental disagreements with the use of animals for food or for pets, and have antipathy towards organised commerce. It is unrealistic to suppose that everything these groups oppose can be eliminated, but there may be alternative structures towards which the agrifood industry could be directed.

Industry asks activist groups what alternatives there are to, on the one hand, the capitalist structure with its attendant issues of market position and protectionism, profits, dividends, investment and indebtedness, and on the other hand, the centrally planned economy, with centrally decided targets, centrally organised marketing and, on all evidence to date, centrally fostered inefficiency, waste and environmental harm.

If we accept that technological advances might be able to make some useful contributions to improved animal, environmental and human well-being, we should ask what tools are best for ensuring an ethical approach to commercial and technical advances: fiscal inducements to acceptable actions; codes of practice designed by committees on ethics; formal legal controls and penalties; or bans on patents?

My view is that a combination of the first three will be more effective for industry and more humanitarian in the short- and long-term than the last, which will irreparably affect intellectual freedom and technological development.

Animal welfare is the keynote of any action. Some welfare legislation now makes specific mention of transgenic animals, and industry should recognise that it is in its own interest to support this legislation fully and indeed to make suggestions to ensure that its scope is more than adequate and its teeth are sharp. Industry and interest groups should work together to ensure that any benefits obtained from transgenic technology are not achieved at the expense of cruelty to animals.

In my view there are a number of unsubstantiated assertions with regards to the patenting of animals, as follows: patents will make it illegal for farmers to breed from GMAs; patents will lead to a drastic decrease in farmers' independence; patents will decrease the number of small farmers; patents will inhibit access to diversity of germplasm; patents will lead to new foods being developed because of their patentability not their quality; patents will undermine public research; patents will increase the power of the biotechnology companies; patents will produce a monopoly control of farming by a few large companies; patents on genes give multinationals control over the genetic resources of the Third World; patenting animals will lead to the destruction of the principles of animal welfare; and patenting will result in the undermining of all respect for nature. The industries responsible for biotechnological advances and for agricultural inputs dispute these unsubstantiated assertions about the patenting of animals. If industry and interest groups start to cooperate then we can avoid all these unsubstantiated assertions about the harm that patenting might cause which currently inflame the debate, distort decision-making and prevent constructive action.

13 Patenting of Transgenic Animals: A Welfare/Rights Perspective

Peter Stevenson

The issue of patenting of genetically engineered animals has become what is known by public relations and marketing people as 'sexy', that is, it is interesting, alive and sells newspapers. This is largely thanks to the 'oncomouse' [see also Chapter 12] and the proposed European Commission's 'patenting' Directive (European Commission 1988d). A patent is a legal right which enables an inventor to require those wishing to use the invention to pay him or her a fee. The principle is that someone who has invested skill, time and money in inventing something should be protected from others using the invention without payment. The essential legal requirement is that you can only get a patent for an invention. Historically, the patenting system has excluded the patenting of plants and animals.

It was never thought until recently that a patent could be granted for an animal. For how could a human possibly 'invent' an animal? The development of genetic engineering has, however, made it possible to modify an animal's genes, and some believe that such an 'invented' animal should be patentable. Others believe that it is contrary to morality, first genetically to engineer animals except in certain limited circumstances, and secondly for patents to be available in respect of such animals.

THE MORALITY DEBATE

The morality debate has to some degree focused on Article 53 (a) of the European Patent Convention (EPC). This article excludes from patentability inventions the publication or exploitation of which would be contrary to public order or morality. People concerned with animal welfare and animal rights have argued that Article 53 (a) means that genetically engineered animals cannot be patented as their exploitation is contrary to morality.

Consideration of Article 53 (a) has largely been focused on the so-called 'oncomouse', a mouse genetically engineered to be highly susceptible to developing cancer. This is the first animal to be patented under the European Patent Convention (EPC). The case is commonly referred to as that of the 'oncomouse' but I wish to stress that the patent in fact extends to any 'onco'-animal, that is, any non-human mammal with the inserted activated 'oncogene' sequence.

In its initial decision on the 'oncomouse', the Examining Division of the European Patent Office (EPO) in Munich refused to grant a patent, although it did so whilst 'side-stepping' the Article 53 (a) morality question. However, the EPO's Technical Board of Appeal resubmitted the case to the EPO's Examining Division for further consideration of certain issues, including the question as to whether Article 53 (a) was a bar to patenting in this case. In its decision the Board of Appeal stated that the decision as to whether or not Article 53 (a) is a bar to patenting the invention in question would seem to depend mainly on a careful weighing up of the suffering of animals and possible risks to the environment on the one hand, and the invention's usefulness to mankind on the other. The key point here is that the Board of Appeal elaborated a test for assessing whether an 'invented' animal is contrary to morality and therefore not patentable.

In May 1992 the Examining Division of the EPO duly reconsidered the case and granted the patent, having come to the conclusion that the usefulness of the 'oncomouse' to humanity outweighs the suffering of the animals [see also Chapter 12].

OPPOSITION TO THE 'ONCOMOUSE' PATENT

The EPC allows for formal 'oppositions' to be made against the granting of a patent, and many European animal welfare groups plan to file 'oppositions' to the 'oncomouse' patent. They will no doubt argue that the usefulness of the 'onco-mouse' has been exaggerated and that it will be of limited value in testing anti-cancer drugs and in testing the carcinogenicity of substances. There is indeed a strong case for arguing that if the Technical Board of Appeal's balancing test had been properly carried out, the Examining Division of the EPO would have concluded that animal suffering outweighed human benefit and that therefore the 'oncomouse' is not patentable. Many animal welfare and animal rights groups believe that the Board's test is not the right way to determine the morality of producing 'onco'-animals, or indeed any genetically engineered animals.

The above case concerned a transgenic laboratory animal. There are also organisations who wish to patent transgenic farm animals. The principal aim of the genetic engineering of farm animals is enhanced productivity. In my view, patented farm animals will be victims of severe stress, since their bodies will be designed, for example, to produce a higher milk yield, or grow faster, bigger or leaner. Indeed, one application before the EPO seeks to patent a transgenic chicken with the bovine (cow) growth hormone gene. The purpose of this invention as stated in the patent application is to obtain faster growth, leaner meat and earlier sperm production in males. The modern broiler has been bred to grow from a chick to a huge 5 lb (2.25 kg) in just 7 weeks, at which age it is slaughtered. Its legs cannot properly carry this massive body. Many birds suffer from painful leg and foot deformities. Many die of disease, including heart attacks and cancer, before the age of 7 weeks. There is clearly a danger that a considerable strain will be imposed upon a transgenic chicken spurred on to yet faster growth by the bovine growth hormone gene. Moreover, such a patented chicken is quite unnecessary. If more poultry meat is needed, the answer is to breed more broilers rather than produce a transgenic superchick.

In Australia a patent application has been made for 'methods of creating new breeds of mammals'. It envisages the creation of transgenic farm animals, including pigs, sheep, goats and horses. Such animals could, for example, be designed to have increased weight gain, feed efficiency or milk production. The patent application gives as an example the creation of a pig with an extra porcine growth hormone gene. Selective breeding has already imposed massive stress on pigs – many have painful joints and some young pigs die of heart attacks. Is it really acceptable to impose yet more pain and suffering on pigs in pursuit of ever faster growth rates?

As I stated above, many animal rights and welfare groups believe that the EPO's suffering-versus-benefit balancing test is not the right way to determine the morality of creating genetically engineered animals. They believe that it is inherently contrary to morality to engineer genetically an animal that is highly likely to suffer. Moreover, if the creation of such an animal is contrary to morality it cannot become morally acceptable to create the animal simply because in a certain case it is judged to be highly useful to humanity to do so. Otherwise our morality ends up taking second place to our sense of what is expedient. Certain values and principles are worthy of our respect even when it turns out to be to our detriment to adhere to them. Looked at from this viewpoint, the EPO's Board of Appeal's suffering/usefulness test is not the right way of assessing the morality of genetically engineering an animal. Furthermore, I shall argue that animals are in any event not suitable subject matter for patenting.

At the heart of the patenting system is the requirement for there to be an 'inventive step'; a patent can only be granted for an invention. In most religions animals are regarded as part of God's creation. From a religious viewpoint it is offensive, even blasphemous, for people to claim to have invented an animal; it is God, not humanity, who creates life (see, for example, Linzey 1990).

From a secular viewpoint, many people accept that we must reassess our relationship to the world in which we live. We must see ourselves as part of that world rather than regarding it as something created for our use. So, too, a growing number of people believe that animals are to be seen not as something placed in the world for our convenience, but as our fellow creatures and capable, like us, of feeling pain and stress. Patenting, regarding, as it does, genetically engineered animals as inventions, as things, is out of step with such modern attitudes to animals.

THE ROLE OF PATENTING IN THE BIO-REVOLUTION

I wish to consider now the question of the importance of patents in the future development of modern biotechnology. Some individuals have asked: why make such a fuss about patenting life forms when genetic engineering will go on even in the absence of patents? Whilst this is true, it is also the case that the availability of patents for genetically engineered products and processes is giving an enormous boost to genetic engineering. The general view of those involved in the biotechnology industry is that companies want the protection of patents if they are to invest huge sums in developing genetically engineered animals.

The importance placed upon patenting for the success of modern biotechnology was highlighted in a document published by the European Commission in April 1991 entitled 'Promoting the competitive environment for the industrial activities based on biotechnology within the Community' (SEC(91)629 final). This document, popularly known as the 'Bangemann Report' after European Commissioner Martin Bangemann, asserts that the biotechnology revolution will ultimately have an impact on our everyday lives as profound as that of information technology. It adds that intellectual property rights (that is, patenting) is a key factor affecting the competitiveness of the bio-industries.

The European Commission, having emphasised the crucial role of patents, goes on to state that they have published a proposal for a Council Directive designed to harmonise EC patent law as regards biotechnological inventions – *Proposal for a Council Directive on the Legal Protection of Biotechnological Inventions COM (88) 496*. The harmonisation of patenting law is described in the Bangemann Report as an essential element in the Community's [European Union's] multifaceted strategies for biotechnology. This proposed Directive is so unthoughtful of the implications of patenting for animals that it affords them no protection at all. Unlike the EPC, it does not even include a morality clause. Fortunately however, the European Parliament voted to insert a clause in the proposed Directive that would exclude inventions from patentability which, 'would offend against public order or common decency'.

The European Parliament adopted two further clauses which give guidance as to the kinds of inventions involving animals which are deemed contrary to public order. One clause excludes chimaeras and animals that cannot be kept without adverse effects on their health. Another excludes inventions causing suffering or physical harm. I do dislike the term *unnecessary* suffering. Who decides what is 'necessary suffering'? When does it become unnecessary? In my view, the phrase 'unnecessary suffering', so beloved of legislators, has done very little to protect farm animals. Nonetheless, the European Parliament's clauses would give some protection to animals, although I would much prefer to see a clause clearly excluding animals from patenting altogether. However, even the limited protection recommended by the European Parliament may not be accepted by the Council of Ministers which, appears to me, invariably to sacrifice the welfare of animals to the perceived needs of industry.

The Department of Trade and Industry (DTI) represents the UK in negotiations on the proposed patenting Directive. The DTI is not sympathetic to the inclusion of clauses in this Directive which would give protection to animals, taking instead the view that animal welfare is not a matter of concern for the patenting system. I think the DTI misjudges the mood of many people in Britain. In 1992, the National Federation of Women's Institutes (NFWI), representing a huge membership, passed a resolution urging that there should be no patenting of animals until all the issues had received thorough public debate.*

In conclusion, I believe that it is quite wrong for patents to be granted for genetically engineered animals. The availability of such patents will give a huge boost to genetic engineering, a technology to which I am opposed except where there is a clear welfare benefit for the animals. The patenting of animals, moreover, runs quite counter to the view of animals as our fellow creatures, and not as items placed on this earth for our convenience.

---

* Editors' update on the European Commission's Proposal for a Patenting Directive: In December 1992 the European Commission modified its proposal for a patenting Directive to take the European Parliament's views into account, and in December 1993 the Trade Ministers of the European Union approved the Commission's revised draft as a 'common position'. In February 1994 the the draft Directive was adopted by the Council of Ministers. However, there were insufficient Members of the European Parliament (MEPs) present to complete parliamentary voting on the draft Directive at the May 1994 Plenary Session in Strasbourg. The Directive then went through the new 'conciliation procedure' where a compromise position was reached between the Parliament and the Council of Ministers in the form of the 'Rothley Report'. On 1 March 1995 the European Parliament voted against the Rothley Report, thereby removing the Directive from the EU's agenda.

Discussion III

Les Ward, Advocates for Animals: I was pleased to hear you say, Mr Lloyd-Evans, that you are opposed to what we would call the extreme manipulation of animals, e.g. the production of chickens without any wings, but I would like to ask you to comment on what seems to me to be a contradiction in your position. You said that the genetic manipulation of animals to make them produce more eggs, meat and milk would reduce animal suffering. To me that would seem to be a contradiction. As Peter Stevenson said, there is already plenty of suffering amongst farm animals so how on earth can you reduce suffering by making these creatures work even harder?

Meredith Lloyd-Evans, MRCVS: I think there are a couple of points here. We all know that conventional breeding has generated Holstein Friesian cows which are champion milk producers, but they happen to have udders which are so big that they constantly stand on them when they are getting up and down from the cubicles, and that if they were allowed to rear their calves, they would never get any milk because they cannot get their heads low enough to put their mouths on the teats. It is possible that through the application of biotechnology one could produce a Holstein Friesian cow which has an udder of a normal size but because of improved nutritional efficiency, physiological efficiency and lactation production efficiency within the cells of that udder, the genetically manipulated cow would produce more milk than a cow with an udder that size produced through conventional breeding practices.

To me, that would be a legitimate aim of developing a transgenic cow. Now you may disagree completely with the idea that cows should be used for producing milk and that milk should be used for human consumption. However, if we get over that hurdle, then I say that biotechnology could be used in a positive way both for the production of food and for the animal's well-being. There are, of course, other animal welfare issues, such as how that animal is handled, and what the conditions on the farm are like, which biotechnology and the patenting of that transgenic cow have absolutely nothing to do with. These other issues have to do with codes of practice for rearing animals and farming systems.

Florianne Koechlin, European Co-ordination, 'No Patents on Life', Switzerland: Mr Lloyd-Evans, did you say that even for industry the patent system is not a good system for the protection of industrial property in comparison with other systems? Could you please clarify this point?

Meredith Lloyd-Evans: The patent system is not the only way that information can be managed by a company or by a person. We all have secrets that we do not tell other people. Similarly, there are companies, particularly in the manufacturing industries where a lot of their ability to provide goods cheaply and efficiently depends on the way they use their technology 'in-house', which do not take out patents for that technology. Instead, they keep the technology secret. Confidential 'know-how' has always coexisted along with the patenting system. The main difference between trade secrets and patents is that the patenting system obliges the patent applicant to reveal the invention in return for protection from other people who try to steal the idea. With confidential 'know-how' you just keep it to yourself and hope that no spy comes and steals it.

The animals themselves are not patented but if patented technology has been used in generating that animal, that bears some sort of royalty. The situation is really no different from when you buy a car or a record player: you do not know who made the different components but at some point some of the the money that you pay gets paid to four or five different royalty bearers.

Sue Mayer, Greenpeace: I wonder if Mr Meredith Lloyd-Evans would agree that the people who are being really innovative are the patent examiners at the EPO. For example, how can we have a situation with explicit exclusions on the granting of patents for animals and plant 'varieties', so that we can patent cows but not Friesian cows? And also, where we have explicit exclusions on biological processes, and yet justify the patenting of subsequent generations?

Meredith Lloyd-Evans: The historical reason for the exclusion of animal and plant 'varieties' is that the EPC was written at a time when the UPOV was in place, so plant varieties (that is, the products of conventional breeding) were already covered, and it was envisaged at that time that a similar system would be created for animals. Now, for whatever reason, that never took place.

A 'variety' of an animal is not something that has been artificially created by manipulation of the genome. A variety of an animal is something that has been created by an ordinary male and an ordinary female mating together, possibly with the aid of *in vitro* fertilisation. As it happens, breeders get a tremendous premium for animal varieties they have bred. For example, in August 1992 a Holstein Friesian cow was sold for £68 000 at a market in the UK. So all this 'malarkey' that the patenting of animals is going to prevent the farmers getting what they want is a load of 'hogwash'. If they have £68 000 they can go and buy a pedigree animal, a 'variety'. They are not going to have to pay £68 000 to get a genetically engineered patented chicken or a piglet or a calf that will grow.

Peter Stevenson, CIWF: I would like to say something concerning animal 'varieties'. Given that the EPC does not permit the granting of a patent for an animal 'variety', you would have thought, as indeed the EPO's own Examining Division thinks, that would mean that you cannot patent a type or kind of an animal. As a lawyer though, I have to point out that the law and common sense very rarely stay together for very long! Lawyers have managed to come up with such a sophisticated definition of what an animal 'variety' is as to make it almost meaningless, and therefore the protection for animal varieties written into the Convention proves to be no protection at all. Sadly, though, one can still get a British minister standing up in the House of Commons, telling Parliament that they need not worry about the Patenting Directive because it protects animal varieties, and most people listening to that statement would think, 'Oh well, that's O.K. Animals can't be patented.' So there is a very sad story behind the term animal 'varieties'.

Meredith Lloyd-Evans: Peter, I think you forget that the Examining Division is only there to examine patents; it is not there to decide points of law. It does not have a locus to decide on whether an animal is an animal 'variety' or not. It can make a decision, but as soon as that decision is appealed against it has to send the matter to a higher patent court for clarification.

Peter Stevenson: Well, with respect, that is not quite the case. The EPO's Board of Appeal said to the Examining Division, 'For goodness sake, get your act together and try and decide what *you* think an animal "variety" is.'

PART IV

Genetic Engineering of Laboratory Animals

14 Rescuing Transgene Expression by Co-Integration

Drs A. J. Clark, A. Cowper, R. Wallace, G. Wright and J. P. Simons

BACKGROUND

Many genes, particularly hybrid genes containing cDNA* sequences, are expressed inefficiently in transgenic animals. The expression of such genes is thought to be strongly influenced by position effects at the chromosomal site of integration. To test whether gene expression can be improved – or 'rescued' – in transgenic mice by manipulating the site of integration, we made a gene hybrid comprising the efficiently expressed sheep beta-lactoglobulin (BLG) gene with two poorly expressed genes encoding human proteins, namely, human alpha-1-antitrypsin (alpha-1-AT) and human Factor IX

---

* Editors' note: Many genes in the genomes of higher organisms (eukaryotes) are interrupted by non-coding regions, known as introns. In contrast to this, the complementary DNA (cDNA) sequences commonly used in genetic engineering do not contain introns. It has been found that the level of expression of cDNA in genetically engineered eukaryotic cells is low. One suggestion is that cDNA is not expressed efficiently because it lacks introns. However, for certain genes there is no practical alternative to the use of cDNA, for example, because the structure of the genomic intron-containing gene is not known, or the DNA sequence containing introns would be too long to be used in genetic manipulation. Another potential factor contributing to the low level of expression is that the cDNA is not being integrated in an appropriate position in the host cell genome. This contribution, which was presented at the Conference by Dr Ron James, of Pharmaceutical Proteins Ltd, describes an experimental approach to improving the expression of foreign cDNA in eukaryotic cells by engineering the site of its integration.

(FIX). In each case, we observed a remarkable improvement in the frequency and levels of expression of the human gene. Our results demonstrate that transgene expression can be rescued by using such hybrids whereby the human gene is co-integrated with the sheep gene, and we suggest that co-integration may provide a general solution for improving the efficiency of gene expression in transgenic animals.

Pharmaceutical Proteins Ltd are developing transgenic animals for the production of valuable proteins in milk. To target expression to the mammary gland, we have used regulatory sequences from the gene encoding BLG – the sheep whey protein. The unmodified BLG gene is expressed at very high levels specifically in the mammary gland of transgenic mice (Simons *et al.* 1987; Ali and Clark 1988; Harris *et al.* 1991). Using sequences from the 5-prime (5') flanking region (the front end) of the BLG gene linked to human alpha-1-AT *genomic* sequences we have obtained high-level expression of human alpha-1-AT in transgenic mice (Archibald *et al.* 1990) and sheep (Wright *et al.* 1991). By contrast, these same 5' BLG sequences linked to an alpha-1-AT cDNA (construct AATD) or a human Factor IX cDNA (construct FIXD) were expressed very inefficiently in transgenic mice (Whitelaw *et al.* 1991).

The inefficient expression of intronless constructs is now well documented (Brinster *et al.* 1988; Whitelaw *et al.* 1991; Palmiter *et al.* 1991). Poorly expressed transgenes appear to be highly influenced by chromosomal position effects. Expression levels may vary widely between transgenic mouse lines, bearing no relationship to the number of copies of the gene construct per cell (copy number), and different lines may display different spatial and temporal patterns of expression (Allen *et al.* 1988; Al-Shawi *et al.* 1990; Bonnerot *et al.* 1990). We reasoned that engineering the site of integration rather than the construct *per se* could provide a strategy to 'rescue' poorly expressed transgenes. For expression in the mammary gland, the vicinity of an actively expressed gene encoding a milk protein could rescue expression of otherwise poorly expressed constructs.

The method chosen for the introduction of a transgene in the vicinity of an endogenous milk protein gene was the co-injection of two genes into the pronuclei of mouse zygotes with the objective that the two genes will co-integrate at a single site (Storb *et al.* 1986; Behringer 1989). Co-injection thus provides a direct method for integrating a poorly expressed transgene in the vicinity of one that is efficiently expressed. To test the rescue hypothesis, transgenic mice were produced by co-injection of the BLG gene with either the AATD or the FIXD construct. Expression was analysed in transgenic mice which carried the BLG gene co-integrated with these second constructs.

CO-INJECTION OF THE BLG GENE WITH AATD OR FIXD CONSTRUCTS

Production and Analysis of BLG/AATD (BAD) Mice

In the first experiment, in which transgenic mice were generated by co-injection of the BLG gene and the AATD hybrid gene in a 1:1 molar ratio, 11 of the 20 resultant transgenic mice carried both BLG and AATD. From these BLG/AATD (BAD) mice, 9 transgenic mouse lines were established. Southern blotting experiments carried out on a number of first-generation (G_1) progeny showed consistent patterns of AATD and BLG-specific fragments within each line. In each line the BLG and AATD transgenes have co-segregated, strongly suggesting the same site of integration. In 1 of the transgenic mouse lines a relatively prominent additional fragment was observed in some of the G_1 mice, most probably resulting from a second site of integration. As expected, copy numbers varied widely between lines (from approximately 1 to in excess of 20), although BLG and AATD copy numbers were similar within each line, reflecting the molar ratio of the DNAs injected.

Milk was collected from transgenic G_0 or G_1 females and human alpha-1-AT was detected by enzyme-linked immunosorbant assay (ELISA) in the milk of mice from 7 of the 9 lines. The concentration of human alpha-1-AT in the milk ranged from approximately 1 microgram per millilitre (μg/ml) to over 600 μg/ml. This differs from expression of the AATD construct in single transgenic mice where expression of human alpha-1-AT, at a low level, was detected in only 1 of 8 independent transgenic mice lines (Whitelaw *et al.* 1991).

RNA was prepared from the mammary gland and liver of BAD transgenic mice and probed for AATD messenger RNA (mRNA) by Northern blotting. An AATD transcript of the predicted size (approximately 1550 nucleotides (nt)) was detected in mammary RNA in 6 of the 9 BAD lines, all of which expressed human alpha-1-AT in milk. The mRNA levels in these mice correlated with the concentration of human alpha-1-AT in the milk. In BAD lines expressing only 1–4 μg/ml of the AATD, transcripts were only just detectable in Northern blots of total RNA, and in the lowest expressing BAD line we failed to detect AATD mRNA transcripts. In mice from the two highest expressing lines, the AATD mRNA levels were estimated by densitometric analysis of autoradiographs to be approximately 4 per cent and 45 per cent, respectively, of the steady-state alpha-1-AT mRNA levels present in a sample of human liver RNA. Further Northern blotting experiments using RNA prepared from a variety of tissues (liver, heart, spleen, salivary gland and kidney) demonstrated that AATD was expressed only in the mammary gland. By contrast, we previously detected no AATD transcripts in mammary gland RNA from any AATD single transgenic mice, including the one expressing mouse in which low levels of the protein were detected in milk (Whitelaw *et al.* 1991).

The frequency and levels of alpha-1-AT RNA and protein expression in BAD transgenic mice contrast markedly with expression of the AATD hybrid gene when integrated alone where low-level expression was observed in only 1 mouse out of 4 G_0 mice and 4 transgenic lines analysed. While caution must be exercised when comparing G_0 transgenic mice with transgenic lines because of potential somatic mosaicism, both the frequency and the levels of expression of the AATD transgene were greatly increased in the BAD transgenic mice. These data clearly demonstrate that expression of AATD was rescued by co-integration with the BLG gene.

Production and Analysis of BLG/FIXD (BIX) Mice

To extend the above observations, a second experiment was performed in which the BLG gene was co-injected with the FIXD construct. In this case, the BLG fragment was injected in a three-fold molar excess over the FIXD fragment, and of 30 transgenic animals obtained, 20 carried both the BLG and FIXD genes. Among 13 of the BLG/FIXD (BIX) G_0 transgenic mice, germ-line transmission was obtained from 11. As was the case with the BAD mice, Southern blotting analysis of a number of G_1 progeny in each of the lines showed the same patterns of BLG and FIXD fragments co-segregating within a line. In contrast to the BAD mice, the BLG copy numbers were consistently higher than those of FIXD, again reflecting the ratio of input DNAs.

Milk was collected and was assayed for the presence of human Factor IX (FIX) by enzyme-linked immunosorbant assay (ELISA) (Wright *et al.* 1991). FIX was detected in milk from 8 out of 12 BIX transgenic mice or lines, although the concentrations were relatively low, ranging from 0.2 to 1 μg/ml. These results contrast markedly with those obtained from FIXD single transgenic animals in which none of 9 mice lines expressed detectable Factor IX (Whitelaw *et al.* 1991).

In the BIX mice, FIXD transcripts were detected in mammary gland RNA samples from 9 of the 11 mice lines analysed. Between lines the steady-state mRNA levels differed considerably, but 3 of the mice lines exhibited high levels of expression, and the FIXD mRNA levels were estimated to be at least 15-fold higher than the steady-state level of human FIX mRNA in a sample of human liver RNA. Further Northern blotting experiments with a variety of BIX lines showed that FIXD expression was limited to the mammary gland. These data contrast with our previous results, in which no FIXD mRNA was detected in any FIXD single transgenic mice (Whitelaw *et al.* 1991). By comparison with a variety of markers including BLG, actin and 18S ribosomal RNA, the FIXD transcripts detected in BIX transgenic mice were estimated to be approximately 1400 nt in length. This is 450 nt shorter than the size expected from the structure of the FIXD construct. These truncated transcripts hybridise to both 5'-specific and 3'-specific (back end) BLG probes (FIXD transcripts are predicted to contain 32 nt and 269 nt of 5' and 3' BLG sequences, respectively) suggesting that the transcripts are deleted *internally*, most probably by an aberrant splicing event. The FIX detection in milk may result from the translation of a small proportion of non-deleted FIXD transcripts, the inefficient translation of the 1400 nt transcripts, or the inefficient secretion of an aberrant FIX encoded by the shortened transcripts. Notwithstanding the low concentration of FIX in milk, it remains clear that the expression of FIXD had been dramatically improved by co-integration with BLG.

Expression of BLG in BAD and BIX mice

The relationship between BLG expression and the expression of co-integrated transgenes was investigated in BAD and BIX transgenic mice. In 6 of 9 BAD transgenic lines, abundant BLG transcripts were detected in the mammary gland, and in these mice high levels of BLG (1 mg/ml) were found in the milk. In each BIX transgenic line investigated, BLG mRNA and protein were detected and, again, a range of expression levels was observed. In neither BAD nor BIX mice was a correlation evident between the levels of expression of the BLG gene and the levels of expression of the AATD or FIXD hybrid genes. In all the mice analysed the expression of BLG was, as expected, limited to the mammary gland.

BLG genes are invariably expressed in single transgenic mice and at levels related to the number of gene copies integrated. In 2 of the 9 BAD lines, however, BLG expression was not observed. This suggests that the AATD construct, in addition to being poorly expressed when integrated alone, can actively *suppress* expression of closely linked BLG genes. Complete suppression of BLG expression was not seen in any of the BIX mice, and this may reflect the excess of BLG sequences in the arrays. Alternatively, FIXD sequences may not have the same negative influence on BLG expression that AATD sequences appear to have.

Composition of transgene arrays

Considerable variation in the levels of expression of AATD or FIXD was observed between different BAD or BIX lines. This may reflect the composition of the transgene arrays integrated in the different mouse lines. A more detailed Southern blotting analysis of BAD lines showed that the co-integrated arrays were often complex and quite different when independent lines of mice were compared. For example, at some loci (places in the genome) some of the copies of the two transgenes are adjacent. However, arrays of BLG were also identified in which the AATD transgene is not present showing that at any one locus there may be a mixture of interspersed and non-interspersed arrays.

It is likely that there will be optimal arrangements of the two transgenes for rescue and we are attempting to construct maps of the arrays in both BAD and BIX mice, although the complexity of some of these arrays makes this a very difficult task. One solution for reducing the complexity of the arrays will be to link the transgenes together prior to micro-injection.

DISCUSSION

The data show that the expression of AATD and FIXD were significantly improved by co-integration with the intact BLG gene. While we cannot exclude a mechanism whereby the BLG gene in some way targets integration of mixed concatamers to permissive chromosomal sites, we feel that this is a remote possibility as there is no evidence for this type of process operating during integration. Rather, we favour a mechanism(s) dependent upon interactions that take place once the transgenic locus has been generated. Three possible models are envisaged. In the first model, BLG genes provide positive effects on the expression of adjacent genes through an enhancer or Locus Control Region (Grosveld *et al.* 1987) (LCR)-like activity. In this model sequences responsible for this activity must be present within the transcription unit of the BLG gene since both AATD and FIXD contain the same 5'- and 3'-flanking segments as the gene.

In the second model it is postulated that the open chromatin conformation associated with the actively expressed BLG gene spreads to encompass adjacent rescued genes. In this model the active expression of the BLG gene creates a permissive domain, for example by allowing the access of soluble transcription factors to the promoter segment of the rescued genes.

In model three the BLG gene contains 'insulating elements' (such as the elements described for the heat-shock gene of *Drosophila)* which serve to shield the transgenes from negative influences of the surrounding chromosomal regions.

Although we cannot formally distinguish between these models at this stage, model three seems the least likely because both FIXD and AATD contain the identical 5'- and 3'-flanking sequences that are present in BLG. Thus, the insulating elements would have to be located within the BLG gene itself and it is difficult to imagine how such elements could function when located in this position.

On balance, we favour model two over model one because

it is known that the BLG transcription unit does not contain essential regulatory elements (Archibald *et al.* 1990; Whitelaw *et al.* 1991) or discretely organised LCR-like elements, as evidenced by the absence of nuclease hypersensitive sites (Whitelaw *et al.* 1993). It will be possible to distinguish experimentally between these two models by determining whether a transcriptionally inactivated BLG gene is competent to rescue.

If transgene rescue can be generalised to other gene combinations and other tissues it will have important practical implications for transgene design. Notwithstanding any potential problems of mRNA processing, the fact that this strategy appears to overcome a requirement for introns for expression means that simple cDNA constructs, like the ones used in this study, can be utilised to overcome the limitations imposed if the structural gene is not available, or is too large for practical manipulation. It is possible that transgene rescue may provide a means of augmenting the expression of poorly expressed constructs that contain introns, especially if they are susceptible to position effects. Finally, the ability to augment the expression of any type of transgene will be particularly important for the production of transgenic livestock, where low efficiency and high cost of introducing gene constructs requires that, once integrated, they function with maximum reliability and efficiency.

Acknowledgements
We thank Frances Anderson for care of the animals and Bruce Whitelaw for helpful discussion.

15 Transgenic Animals as Disease Models

Dr Elaine Dzierzak

The uses of trangenic animals for the study of genetic disease and in the development of therapies for human genetic disease are considered in this chapter. Mouse models of human genetic disease have been used for many years. They can be useful for mapping genes related to gene function and regulation. Animal models are also useful for testing novel therapies and in the study of embryonic and foetal development (which is not considered here).

MUTATIONS

Genetic mutations occur spontaneously both in wild mice and in mice that have been inbred in the laboratory. Many years ago, mouse breeders in the UK, in the USA and in the Far East, noticing that there were physical differences between mice, deliberately bred from those strains with physical defects. In the first instance such mice were bred for characteristics like coat colour differences, curly tails or short ears, but some mutant strains of mice and other laboratory animals became useful in the study of genetic diseases. For example, there is a strain of rat that has aided us tremendously in the study of familial hypercholesterolaemia so that we now know it is the low-density lipoprotein (LDL) receptor that is responsible for the high cholesterol levels characteristic of this inherited condition. Various mouse mutations, and even mutations in the pedigree dog population, which are associated with genetic deficiencies, can be used to study genetic disease.

I would like briefly to describe a naturally occurring mutation in inbred mice which has proved useful to medical research. It is called the 'shiver mutation'. A mouse with this mutation has tremors and convulsions which are due to the absence of a protein covering of the nerve cells. The protein, called myelin basic protein G, both insulates nerve cells and aids in the transmission of nerve signals down the nerve fibres. When exogenous protein G was microinjected into fertilised mouse eggs of the shivering mouse strain the resultant mice no longer shivered. This constituted the first proof that the myelin basic protein G is the defective protein in the shivering mouse strain.

METHODS

Through the use of such natural mutations in mice we have been able to discover quite a lot about human genetic disease. To further these studies we have now moved into the field of genetic engineering. There are a number of methods by which we can introduce new genes and gene mutations into inbred mouse strains for the study of human genetic disease. Three methods of transgenesis are outlined here. One method is microinjection which introduces a new gene into a random location within the genome of the host cell. A second method is homologous recombination which is the targeting of a foreign gene to a specific position in the host cell's chromosomal DNA. The third method of transgenesis is vector mediated which is usually undertaken using viruses as vectors to transfer foreign genes into certain cells within the host animal.

Microinjection

Microinjection is undertaken when the fertilised egg is still in the single-cell stage. The second method of transgenesis – homologous recombination – utilises the blastocyst stage. Thus the first two methods of transgenesis produce animals that can pass on the transgenic trait when they breed. The third method is used only to transmit genes into somatic (body) tissues of an individual. The resultant transgenic animal can never transmit that transgene to its offspring.

Microinjection involves using a little glass pipette to introduce a DNA solution into the nucleus of the fertilised mouse egg. Many transgenic mice have been produced in this way; for example, a lot of mice with foreign beta-globin genes were made because there are many genetic deficiencies in the human beta-globin gene that exhibit themselves as thalassaemias and in sickle cell anaemia (see later). Many genes of the human immune system have also been transferred into trangenic mice to study how the immune system distinguishes between 'self' and 'non-self'. Also, genes of the histocompatibility complex were transferred into transgenic mice by microinjection for studies on grafting and graft rejection.

Homologous Recombination

The second method of transgenesis – homologous recombination – is also known as gene targeting, because the foreign sequence that you want to insert will be inserted in that part of the host genome that contains the equivalent endogenous (host) gene. Where a defective gene is replaced with a normal gene by this method it is called 'gene inactivation'. You could also use this technique in a mouse that has a mutated gene or a gene that is deleted in some way, and it can be used for a regulatory element or a coding region of a gene. Thus, this technique allows a form of 'gene correction' where you could replace defective genes with normal ones.

In order to perform homologous recombination, you take 3.5-day-old blastocysts and culture cells from them, thus creating a tissue culture cell line that can be maintained, genetically manipulated and, in the end, used to produce an animal all of whose cells are derived from the genetically manipulated tissue culture cells. Many transgenic mice have been produced using this technique and have been very useful for studies of genetic diseases and cancer.

Vector Mediators

Finally, the third method of transgenesis is by vector mediators. For example, you can use an RNA virus (retrovirus) to construct a retroviral vector that carries the foreign gene inserted into a region of the viral genome that is not essential for the virus. The virus used for this kind of vector system is debilitated so that it can only infect a cell once, after which it cannot produce infectious virus progeny. Therefore this is a very safe way of introducing a gene into, for example, mouse somatic tissue. Somatic tissues that have been infected with retroviral vectors include muscle tissue, epithelial tissue, tissues of the blood system and cells in the brain.

MODELS

Some examples follow of where transgenic models have been used for the study of first, genetic disease, and secondly, cancer. For genetic disease, the reticular and epithelial system, the blood system and the respiratory system are considered.

Reticular and Epithelial System

In the reticular and epithelial system, a transgenic mouse model has been developed for Gaucher's disease. Gaucher's disease is a lysosomal storage disease of high prevalence in the Ashkenazi population as well as in a number of other populations. It is caused by a deficiency in an enzyme called glucosidase. Glucosidase is responsible for degrading certain kinds of protein that, when they accumulate in the lysosomes of macrophage cells, result in symptoms such as lethargy, nervous system dysfunction and an accumulation of liquid deposits in the liver and in the bone marrow. There is no naturally occurring mouse model for Gaucher's disease and so, in order to develop treatments for this disease, it was very important to create a mutant mouse model. A group of researchers did so using the homologous recombination method. They cloned a mutation into the glucosidase gene and introduced the foreign gene into the glucosidase gene of a mouse. Since the resulting animals mimic the human disease, the mouse model can be used to develop therapies for Gaucher's disease.

Blood System

Red blood cells are very important for carrying oxygen, and haemoglobin is the protein responsible for this. A number of different genes encode haemoglobin, but the beta-globin gene is the most important one in adult red blood cells. Numerous different mutations occur in the beta-globin gene resulting in illnesses such as thalassaemia, which in very severe cases can lead to death at the age of about 13, and sickle cell anaemia, for which a transgenic mouse model has been developed.

Sickle cell anaemia is caused by substitution of one amino acid – glutamate – for another – valine – in the chain of amino acids that constitutes beta-globin. As a result of the altered beta-globin chain, under certain conditions the patient's red blood cells become irreversibly 'sickled' in shape. Such sickled cells block the patient's arteries and can lead to very serious problems in kidney function later in life. Once again there was no naturally occurring model for sickle cell anaemia in the mouse population with which to study the factors that lead to the sickling of red blood cells in patients.

Several years ago a group of researchers in the UK developed a transgenic model for sickle cell anaemia. They took a human gene which contained the sickle cell mutation, microinjected it into fertilised mouse eggs, and identified those mice in which nearly 100 per cent of the red blood cells can be sickled *in vitro*, thus mimicking the human disease. The transgenic mice differ a little from humans with the disease because the normal mouse beta-globin is still present, and so the mice do not suffer in the way that a human with this disease would.

One could potentially cure such a genetic deficiency of the beta-globin gene by adding a normal gene for beta-globin through vector-mediated transgenesis. One could take a thalassaemic animal, remove its bone marrow (bone marrow contains the cells that produce the mature red blood cells), genetically manipulate the beta-globin gene of the bone marrow using vector-mediated transgenesis, replace the bone marrow in the animal, and in this way cure the thalassaemic condition.

Respiratory system

Another model for genetic disease that has been developed is a mouse mutant for cystic fibrosis. Cystic fibrosis is a lethal, recessive genetic disease that affects about 1 out of 2500 newborns. The frequency of carriers in the population is about 5 per cent, so it is a prevalent disease. It affects the respiratory system and the digestive system resulting in obstruction of the airways by mucus and in intestinal obstructions. Children with cystic fibrosis are very underweight.

Again, there is no natural mutation in the mouse population for cystic fibrosis, so a number of laboratories took on the task of developing an *in vivo* mouse model for the disease. When the gene responsible for cystic fibrosis is introduced into mice it affects their size as well as their lungs, in which a lot of mucus accumulates. The animals also show intestinal obstructions so such animals will be very useful in the long run for developing new therapies for the treatment of this devastating disease [*c.f.* Chapter 16].

Cancer Models

Models for chronic myeloid leukaemia (CML) and for brain tumours are briefly described next. CML is a disease characterised by a long chronic phase which at any time may dramatically change into an acute myeloblastic leukaemia. Many scientists are trying to develop new therapies to limit onset of this potentially lethal type of leukaemia. One approach to studying the development of this disease is to create animal models using a vector-mediated transgenesis system to introduce the gene believed to be responsible for CML into laboratory animals. In human patients CML is always associated with a translocation of part of chromosome 9 to chromosome 22. The gene product of the fused regions of chromosome 9 and 22 was believed to be responsible for the development of CML. In order to test this hypothesis, the fusion gene was introduced into the bone marrow of animals by vector-mediated transgenesis to see whether these animals developed the symptoms of CML.

In human patients, CML is accompanied by an enlargement and alteration of the 'architecture' of the spleen. In a mouse whose bone marrow had been infected with the fusion transgene, the spleen was found to be very pale, and to resemble the architecture of the spleen of a human patient with CML. So, once again, the genetically engineered animal model will be useful in studying the mechanisms involved in the progression of the chronic phase of this disease and will provide a system for testing therapeutic regimes.

The final model provides a dramatic demonstration of what transgenesis can do in the case of cancer. This is a study of the use of a retroviral vector for *in vivo* gene transfer to treat experimental brain tumours in rats. The rat subjects develop massive brain tumours over a period of 30 days. Researchers wanted to see if they could introduce a gene which encoded a protein product that was toxic to tumour cells. In other words, could they kill the tumour cells by introducing a tumour toxin gene into them? As a safeguard, the foreign gene only became toxic if a certain drug was administered to the rats. When tested against control animals over a 30-day period, it was found that rats transfected with the toxin gene and given the drug for it to become toxic showed no growth in tumour volume, whereas the control animals showed continuous tumour growth over the 30 days. It was found that introduction of the gene into an animal with a highly developed brain tumour could result in the complete elimination of the tumour cells. This experiment demonstrates the power of such genetic manipulation. It remains to be seen what will be achieved using it in the future.

SUMMARY

Transgenic animal models have aided us in the understanding of human genetic disease and also of cancer and are absolutely necessary for the development of clinical therapies. They enable scientists to study more precisely the effects of specific treatment regimes because transgenesis limits the number of variables in the animal subjects.

The animal experiments that we do for medical research in the UK are controlled by Government-specified guidelines and regulations. People doing these sorts of experiments are licensed at two levels. First the project has to be approved through a governing body for its acceptability and necessity. In addition the individual is licensed for the handling and manipulation of experimental animals. This is under the Animals (Scientific Procedures) Act 1986.

16 A Critical View of the Use of Genetically Engineered Animals in the Laboratory

Dr Gill Langley

In the laboratory, genetically manipulated animals are used to study biological processes, especially those with medical relevance such as the immune system, the regulation of growth and development, genetic disease, cancer and infectious illness. Genetic manipulation can have wide-ranging harmful effects on individual animals, partly because as a process it is not fully controlled, and partly because changing or inserting even a single gene can have dramatic and sometimes unforeseen results.

Genetic engineering is often presented in an unjustifiably positive light. To redress the balance a little, this critique of genetic engineering has two main thrusts: the scientific inadequacies of the technology which contrast starkly with the optimism of genetic researchers; and the unacceptable suffering caused to animals both by the unpredictability of the manipulations and by the inherent disruptive power of genetic engineering. The examples given will hopefully illustrate these linked themes.

SOURCES OF UNPREDICTABILITY

Inserting a transgene into an embryo is only the start. To be functional a gene must be expressed, that is, produce the protein molecule it codes for. Many complex processes are involved in gene expression, and most of these crucial events cannot be controlled by scientists. Here are seven causes of unpredictability, each of which has the potential to harm animals:

- site of gene insertion;
- insertional mutations;
- heritability of inserted genes;
- tissue-specificity of gene expression;
- expected tissue specificity but unexpected results;
- variable gene effects within one organ;
- unexpected reversal of transgene activity.

Site of Gene Insertion

The exact place on an animal's chromosome where a transgene inserts is fairly random, and from one to several hundred copies of the gene may insert in a row at a single location on the chromosome. Different insertion sites dramatically affect the transgene through the influence of the animal's own genes nearby. Promoters, enhancers and silencers, plus additional remote control elements (such as dominant control regions), may determine whether the transgene is expressed, and to what extent.

For example you might expect that the level of expression of a transgene in various tissues would parallel that of the animal's own gene, if they code for the same product, but this is not always so. In normal animals the metallothionein gene is expressed in different tissues and at different levels, and among founder [transgenic] animals some individuals have abnormally high expression in the kidneys or pancreas and almost none in liver. This is thought to be because the metallothionein gene is very sensitive to chromosomal context, and thus is dependent on the site of integration, which will vary among different founder animals.

Insertional Mutations

If the transgene integrates in a way that disrupts an animal's natural genes, an insertional mutation can occur. It is not known how common these are because some kill the embryo at an early stage and others may not be recognised easily if, for example, the effect on the animal is internal.

An illustration of an insertional mutation concerns a transgene, consisting of a fruitfly (heat-shock protein) gene and a viral (thymidine kinase) gene being inserted into mice (McNeish *et al.* 1988). An insertional mutation led to some mice being born with developmental abnormalities including a virtual loss of hindlimbs, malformed front limbs, facial clefts and massive brain defects (missing olfactory lobes, brain haemorrhages, and hydrocephalus). Even though this mutation was unplanned, the researchers felt it interesting enough to produce 72 mutant individuals.

Heritability of Inserted Genes

Variations in transgene integration or activity between founder animals and their offspring complicate efforts to develop animals who breed true, and requires the manipulation and breeding of sometimes hundreds of animals to develop the desired line. Many animals may suffer abnormalities during this lengthy procedure.

If injected genes do not insert immediately into the chromosomes of the single-celled embryo, the animal develops with the gene in some tissues but not in others. This undesired result can occur up to a third of the time, and if the gene is not in the germ cells then it will not be transmitted to offspring. Another complication is that the foreign DNA may become chemically altered (methylated) in offspring, producing variable and unforeseen effects, and some transgenes may not integrate into the chromosome at all but persist outside it, causing bizarre patterns of inheritance.

Tissue-Specificity of Gene Expression

By parallels with normal genes, some inserted genes are expected to be expressed in particular organs and not in others, but it does not always happen that way. For example, a transgene consisting of the hypothalamic gene for the hormone vasopressin and a viral (SV40) oncogene (cancer-related gene) was inserted into mice. The vasopressin gene should have activated the oncogene in the hypothalamus of the brain, causing tumours there. In fact, the mice developed tumours in the pituitary gland and the pancreas but not in the hypothalamus (Murphy *et al.* 1987).

There are species differences in tissue-specificity too. For example, the normal mouse gene for transferrin is inactive in the kidneys, but when inserted into mice the chicken transferrin gene *is* expressed in the kidneys (Palmiter and Brinster

1986). Possibly different species have evolved different signals for tissue-specificity of gene expression, which could make it difficult to extend results from an animal to a human.

Expected Tissue-Specificity, but Unexpected Results

Even if a transgene is expressed in the expected tissues, unpredictable effects can still occur. The normal protamine gene is thought to be expressed specifically in male mice during sperm production. A transgene made of a protamine gene and a viral cancer gene (SV40 T antigen) was expressed in the testes of genetically engineered animals, but there were no testicular tumours. The mice did, however, unexpectedly develop tumours in the heart and bones (Behringer *et al.* 1988).

Variable Gene Effects Within One Organ

Inserted genes may produce their products equally throughout an organ in one individual, and yet have vastly different effects on different parts of that organ. For example, transgenic mice were developed containing a cancer gene (SV40) plus a promoter (atrial natriuretic factor) which would activate the oncogene in the upper chambers of the heart. All the mice expressed the inserted genes *equally* in both the upper and lower chambers of the heart (Field 1988). However, after birth and over a period of weeks the right upper chamber of the animals' hearts grew massively, developing to hundreds of times its normal size and overwhelming the rest of the heart. The left upper chamber was normal, even though it expressed the same level of the same gene. The huge heart tumours caused cardiac arrhythmias and killed the animals.

The cause of the asymmetric response was probably the differential activity of growth factors in the two heart chambers, which made the right one responsive but not the left. That this can occur within one organ of an animal indicates the complex problems to be overcome when scientists attempt to extrapolate across the species barrier from mice to humans.

Unexpected Reversal of Transgene Activity

Some transgenes have been found to be very active when first microinjected into the embryo but unexpectedly were not expressed in the adult animal. The assumption is that they were inactivated at some point during growth and development; such effects are also likely to vary from species to species.

Following integration in the host cell genome, the DNA of inserted genes may undergo rearrangement during subsequent cell divisions, which affects their function. Transgenic mice were developed carrying a gene for the plasminogen activator. All the newborn mice had high levels of plasminogen activator in their blood which, as a result, did not clot. Half the mice had spontaneous bleeding in the abdomen and intestine and died within 4 days. However, the surviving mice gradually reverted to normal levels of plasminogen activator and normal blood clotting. It was concluded that the foreign DNA had undergone rearrangement in some liver cells, and these normal cells had actively repopulated the liver (Anon 1992).

ANIMAL SUFFERING

Genetic manipulation affects the most fundamental but immensely complex processes of cells and tissues. These processes are so poorly understood and controlled that genetic manipulation produces physical abnormalities which can cause pain and distress to many animals. The numbers of animals 'created' during the hit-and-miss efforts to produce a particular desired transgenic individual often run into hundreds, and many of these may suffer severe or lethal defects. Below, just two examples from the scientific literature are discussed: choroid brain tumours and lens tumours.

Choroid Brain Tumours

One of the particular problems with using oncogenes is that, because tumours appear in unexpected places, they may not be detected before they cause substantial pain and distress.

In experiments reported in 1984 (using an SV40 gene in a number of constructs), American scientists were surprised to find some transgenic mice dying by 3 months of age (Brinster *et al.* 1984). When they 'carefully monitored' the mice, they spotted an animal with a bulging skull; when they killed it they found a large tumour in the brain.

Subsequently, they identified several other mice with the same bulging skull, and also noticed occasional signs of brain damage including loss of balance, partial paralysis and uncontrollable twisting of the head. At postmortem, when the skulls of these mice were opened, large amounts of fluid escaped and the brain collapsed. Sometimes extensive brain haemorrhage had obviously occurred, and large brain tumours were seen. Some of the mice had other abnormalities such as kidney damage and massive thymus glands which caused them difficulties in breathing.

Many of the mice died from their brain tumours. This was lamented by the scientists, but only because: 'Some potentially interesting lines were lost because animals with tumours frequently died before progeny could be produced' (see Brinster *et al.* 1984). The researchers reported that a bulging skull was generally the first observable sign that one of these animals had developed a tumour. This means that by the time it was detected the cancer was already large enough to distort the bones of the skull, and presumably to cause suffering.

Lens Tumours

A naturally occurring tumour of the lens of the eye has never been reported. American scientists, wanting to explore whether the lens is absolutely resistant to tumours, produced transgenic mice (with an SV40 oncogene plus a gene for crystallin, the principal constituent of the lens) who did develop lens cancers (Mahon *et al.* 1987). Most of the founder animals died before they were 4 months old, but one line was successfully bred and at the time of writing in 1987, 5 generations of these mice had been produced.

At birth the lenses of the transgenic animals were white and cloudy. By the age of 2 months, the lenses had been replaced with disorganised masses of tissue which burst out of the lens capsules. In fact the tumours continued to grow right through the back of the eye or out through the cornea at the front of the eye. Some mice developed other tumours as well. The scientists expressed the view that because the lens is not an 'essential tissue for survival', it can be removed from living mice 'without losing valuable breeding animals' (see Mahon *et al.* 1987).

Escalating Numbers

Considering the nature and extent of suffering caused to transgenic laboratory animals, it is of special concern that the production and use of transgenic animals are escalating rapidly as genetic engineering becomes the research 'flavour of the day'. In Switzerland the number of research projects using transgenic animals increased 12-fold between 1986 and 1990 (Went and Stranzinger 1990).

In the UK, transgenic animals have only been identified in the official statistics since 1990, when 48 255 animals were recorded. A year later the figure had already climbed to 62 445, a rise of 29 per cent (Home Office 1992). This contributed to the overall increase in registered 'animal procedures' in the UK in 1991, the first documented rise since the mid-1970s.

TRANSGENIC ANIMAL MODELS OF HUMAN DISEASE – TWO CASE STUDIES

Cystic Fibrosis – The 'Knockout' Mouse

In August 1992 the cystic fibrosis 'knockout' mouse was announced to the scientific press (Clark *et al.* 1992; Snouwaert *et al.* 1992). The result of manipulating and breeding hundreds of animals, this transgenic mouse has a deficient protein in the membranes of its epithelial cells, causing a cell defect similar to that suffered by people with cystic fibrosis. However, despite the similarity at the genetic and cellular levels, the cystic fibrosis 'knockout' mouse's symptoms mimic human ones only very imperfectly (Collins and Wilson 1992) [*c.f.* Chapter 15].

Only 5–10 per cent of human infants born with cystic fibrosis suffer from intestinal blockage, but all the cystic fibrosis 'knockout' mice who have been autopsied so far have died of this condition within 30 days. Death was preceded by weight loss, abdominal distension and awkward movement, and was caused by peritonitis and rupture of the intestines and gall bladder.

Eighty-five per cent of cystic fibrosis patients suffer from pancreas problems, but so far very few of the cystic fibrosis 'knockout' mice have shown any signs of this. The pancreas of the mouse is structurally different from that of humans, and even the tissue changes seen in the transgenic mice may have been secondary – due to ill health – rather than a primary result of the gene manipulation.

Lung infections cause 95 per cent of deaths and disability in cystic fibrosis patients, but the cystic fibrosis 'knockout' mice show few signs of airways abnormalities. The researchers hope that the mice, which are born and reared in germ-free conditions, may develop signs of respiratory disease when they are exposed to infection. However, mucus-producing cells and glands in rodents are far fewer than in human airways, another confounding species difference which may limit the value of the mouse model.

Ninety-five per cent of (male) cystic fibrosis patients are infertile. However, no major abnormalities have been seen in male mice, and indeed two of them have successfully impregnated females.

An important symptom in cystic fibrosis patients is abnormalities of the salivary glands. This is not a feature of the cystic fibrosis mouse.

Up to 43 per cent of cystic fibrosis patients develop cirrhosis of the liver, but so far no abnormal signs have been found in the transgenic mouse livers. The researchers hope that if the animals can be made to survive longer, perhaps such changes will develop.

The tissues of people with cystic fibrosis have abnormal electrical activity because the chloride channels in their cell membranes are defective. This was partly mimicked by the cystic fibrosis 'knockout' mouse, but experiments with a drug called amiloride, which from clinical tests in patients shows promise as a cystic fibrosis treatment, revealed a different response in the cystic fibrosis mice: unexpectedly, the mice were found to have a previously unknown membrane channel, not found in humans.

Thus, despite the usual massive media acclaim, the 'knockout' mouse falls far short of original expectations for a model of human cystic fibrosis. There seems no doubt that species differences, both in gene expression and in general anatomy and physiology, will limit the value of transgenic mice in cystic fibrosis research. They may prove useful to study the cellular basis of cystic fibrosis (which has in any case been studied in cell culture), and the intestinal aspect of the disease, but their usefulness for the study of the major clinical symptoms of human cystic fibrosis and the development of appropriate therapies is likely to be limited.

The HPRT Mouse and Lesch-Nyhan Syndrome

In 1987, two British research teams simultaneously published reports that they had, by slightly different techniques, produced mice with a defect in the gene which codes for the enzyme hypoxanthine-guanine phosphoribosyl transferase (HPRT). The gene defect gave rise to the expected lack of HPRT activity (Hooper *et al.* 1987; Kuehn *et al.* 1987).

In humans, deficiency in HPRT causes Lesch–Nyhan syndrome, a rare inherited neurological and behavioural disorder affecting males. Its effects are severe: mental retardation, spastic cerebral palsy, uncontrollable muscle twitching, and compulsive self-mutilation by biting. Both teams of researchers expected that the HPRT-deficient mice would provide a model to study Lesch–Nyhan syndrome. The prospects envisaged included: evaluating drug or transplant treatments; elucidating the defects in brain pathways which underlie the syndrome; developing pre-implantation diagnosis of the human syndrome and gene therapy for affected embryos. However, this particular 'brave new world' was not to be because, despite monitoring the transgenic mice carefully for symptoms, it became clear that although the genetic defect mimicked the human one, it caused no detectable effects in mice at all because they had another enzyme which could perform the same function as HPRT (Barinaga 1992).

IMPERFECT TECHNOLOGY

In conclusion, genetic manipulation is an ill-understood and imperfectly performed technology, and scientists are using living animals as their 'crucibles' to perfect it [see Wheale and McNally 1988, Chapter 7].

Even if genetic engineering can be better controlled, the species barrier remains and will limit the relevance to humans of transgenic disease models. An early lesson in species differences was given when genes for promoting growth were inserted into mice in the 1980s. They grew rapidly, as expected, and were twice the size of normal mice. When the same technique was applied to the 'Beltsville pigs', they suffered a wide and unpredicted range of problems: lethargy, kidney damage, defective vision arising from abnormal skull growth, reproductive abnormalities, susceptibility to pneumonia and other infections, and bone and joint disorders.

Genetic manipulation has created mice whose genetic defect mimics that of the human counterpart, but increasingly it is found that the same genetic defect does not necessarily yield a comparable illness. Since so many interacting factors are involved in growth, development and disease – including environmental aspects – there will always be severe limitations to what genetic engineering can tell us.

Because genetic engineering is productive of unforeseen results, the potential to harm animals is huge, as I have emphasised. Although mice are by far the favoured laboratory animal for genetic manipulations at present, their drawbacks as models of human processes will become more apparent as time goes on. Already, geneticists have concluded that experiments with transgenic mice cannot be used as a substitute for experiments on farm animals because of species differences.

We are faced then with the prospect that species closer to us in general evolutionary terms, or species which resemble humans in particular metabolic or anatomical ways, may be genetically engineered. How will we feel about hundreds of transgenic dogs and monkeys being genetically manipulated to develop tumours or painful chronic diseases? Public opposition to genetic engineering of farm animals, creatures familiar to us from childhood, is already greater than opposition to the use of mice. We can expect levels of concern to escalate to enormous heights when cats, dogs and primates become involved.

The dream of genetic manipulation is the same one that scientists once had for drug therapy – that a magic bullet really could be developed, a tool so precisely and specifically crafted that it would go unerringly to the heart of the disease process, with no shrapnel flying around to complicate matters. We have learned that all drugs, no matter how well targeted, have side-effects because they cannot be entirely specific to the diseased tissue. I believe that we will learn that the dream of genetic engineering is similarly flawed, and the horrible suffering of hundreds of thousands of animals will turn the dream into a nightmare.

Discussion IV

Anita Idel, German Veterinarian: I have a question for Dr James. Could you please comment on experiments on transgenic mice engineered to produce transgenic alpha-1-antitrypsin (alpha-1-AT)?

Dr Ron James, Managing Director, Pharmaceutical Proteins Ltd, Edinburgh: At least two transgenic mice expressed alpha-1-AT in the salivary gland, which meant they probably digested it and it was consequently an additional source of dietary protein! You have probably also ingested alpha-1-AT, because the first thing you do if you cut your finger is to stick it in your mouth, and almost certainly in that blood there is human alpha-1-AT. It does not hurt you when you digest it and it did not hurt the mice either.

We cannot find any expression of alpha-1-AT in the salivary glands of our transgenic sheep. We are not quite sure of the reason why mice express alpha-1-AT in the salivary gland but we think it is a function of the alpha-1-AT gene itself, rather than the regulatory sequence, because we have expressed a number of other proteins in the milk of animals, and so have other people, without seeing expression in the salivary gland. So it is something that is related to the alpha-1-AT gene and not to the regulatory sequence. It does not seem to be causing any harm to the mice, and we have not yet found it in the sheep.

Richard Gard, Institute of Agricultural Medicine, Plymouth University: Can I ask Dr James about the work at Pharmaceutical Proteins Ltd, Edinburgh? Has the work progressed to trial stage?

Dr Ron James: The simple answer to that question is that we would expect to start clinical trials towards the end of 1994. Before we can start clinical trials we have to make absolutely certain that the animals that we are using are disease-free, and that viral clearance studies demonstrate that the purification process is adequate.

Geraldine O'Brien, Cork Environmental Alliance, Ireland: Could I ask Dr Langley her opinion on the work on transgenic sheep by Clark *et al.*, as described by Dr James?

Dr Gill Langley, Dr Hadwen Trust for Humane Research: I was impressed by Dr James' obvious sincerity in expressing concern about the welfare of the transgenic sheep he uses for the production of medical proteins, but I think that we have to take a wider look at the use of animals in the production of proteins in this way.

The presentation we heard made no mention of the fundamental research which underlay his successful production of protein: what animal suffering was involved in the early experimentation in the development of the techniques?

I have another question for Dr James. We have been reassured that the sheep used for the production of pharmaceutical proteins do not undergo any unnecessary suffering. However, I was concerned about what happens when a protein produced using transgenesis is found to be pure and active; will it still be tested in animals for toxicity and efficacy before entering the market? I have a suspicion that it will, because, for example, genetically engineered human insulin had to undergo those kinds of tests. If this is the case, the production of transgenic pharmaceuticals will be followed by further animal suffering.

Dr Ron James: I am not responsible for the answer to this question. It is a question that will ultimately be answered by the regulatory authorities. In the case of transgenic alpha-1-AT, alpha-1-AT which is derived from human blood is already in clinical use, and I am hopeful that if we can demonstrate chemical identity between transgenic alpha-1-AT and the alpha-1-AT which is derived from human blood, we may be able to avoid animal testing. But if we cannot and we have to undertake animal testing, that is a function of what society asks us to do before we put a new product into humans, and not a function of the method that we are using in order to produce it.

I would like Dr Langley not to be quite so dogmatic in her arguments but to be a bit fairer. Now, if somebody asks me, as a scientist, 'Could this cause harm?' If I cannot be 100 per cent sure that it will not, then I honestly say, 'Yes, it could, but maybe, and probably, it won't'. There were an awful lot of 'could be's' and 'maybe's' in your presentation. Some of the examples of transgenic animals that you used clearly did suffer. However, it is not fair to say that *all* transgenic animals will suffer because some did, and it is not necessarily legitimate to argue on the basis that the right side of the heart and the left side of the heart in mice respond differently that therefore a mouse is different from a human. It just does not follow. You have some perfectly logical arguments but let us have a debate which is based on fact and not based on wildly irrational assumptions.

Dr Gill Langley: I am absolutely astonished at that attack on my contribution. For example, I do not say that the differential effects on the heart proved that you cannot extrapolate results. I said that if there are differences in developmental processes and in growth factors within one organ of one animal, then how much more variety is there likely to be between species.

Malcolm Eames, BUAV: I would like to address a question to Dr Ron James of Pharmaceutical Proteins Ltd. You said that you like to deal in facts and so I would like to know when you will feel able to publish details of the morbidity and mortality of all of the animals born under each and every one of your firm's transgenic animal programmes.

Dr Ron James: It is an enormous number but so far as the transgenic sheep are concerned, the vast majority of them are still living. As far as the number of deaths that we get in lambing, they are very much less than the number of deaths that would occur on an ordinary sheep farm.

Jake Wilson: A member of Dr Robert Gallo's research team at the National Institutes of Health (NIH) in the USA has raised the possibility that AIDS research could potentially produce an infectious, possibly virulent, form of human immunodeficiency virus (HIV). Would any of the panel care to comment on the possibility of this, and the risks, given the existence of human error?

Dr Elaine Dzierzak, National Institute for Medical Research: I can only comment on what I know of the places in the USA where they are doing this research. The animal facilities are classed as security category P3, which means they are operated under strict containment conditions. Very few people go in and out of such facilities. The people doing such research do not want to be infected themselves, and will do what they can to protect themselves and the animals in such facilities.

Annabel Holt, Green Network: I would like to ask what the Government should be doing to encourage the development of human tissue culture. For me, to be told I am like a mouse or a rat is absolutely ludicrous and I would much rather the Government were trying to develop tissue banks for humans. I carry a donor card and favour the idea of donating my tissue to replace that of animals in medical research.

Dr Jacqueline Southee, Microbiological Associates International Ltd: There are moves to develop human tissue banks at the moment but we still have to research into how we can preserve that tissue – whether we are going to freeze it, freeze-dry it or just try and maintain it in culture medium. You say you would like to donate your tissues to replace animal experiments. That is a possibility, but the scientists who want to use human tissue to develop alternative systems for medical research are in competition for that tissue with those who want to recover your organs to put into people who need them, for example, people with kidney disease or liver disease. However, it is possible that areas of compromise between the two groups will be found.

One of the problems with using human tissue for the testing of new medicines is its variability; if we use a bit of your liver and a bit of somebody else's liver we get different results because you are different people. Our aim is to get an alternative regulatory system accepted for toxicity and for that, ideally we need to establish a standard tissue system, which is why we are considering the immortalisation of tissue lines. Another problem is that primary tissues quickly differentiate into other tissue types.

At present the testing of medicines in *in vitro* systems is much more expensive than testing in traditional animal systems. One thing that I would like the Government to do is to subsidise research into the development of alternative testing systems through a levy on animal experiments.

PART V

What Place has Genetic Engineering of Animals in Society

17 Genetic Engineering: Can Unnatural Kinds be Wronged?

Professor Robin Attfield

SOME ETHICAL OBJECTIONS TO ANIMAL GENETIC ENGINEERING

What, if anything, is wrong with making genetically new strains of plants or animals, such as 'Factor IX' sheep or 'Beltsville pigs'? (see Bulfield 1990). One view which deserves to be taken seriously is that of Michael Fox, who holds that transgenic manipulation is wrong because it violates the genetic integrity, or *telos*, of organisms or species (Fox 1990). In his view, the natural end, role and niche – or *telos* – should be respected, and thus altering it is justified only in exceptional cases, of which Factor IX sheep may be one. The related view, also referenced by Macer (1989), and taken by Hans Jonas (1984) is that various kinds of genetic engineering are 'unnatural' and thus wrong.

Let us set aside that most of the various species of domestic animals could also be regarded as 'unnatural', and note that both of these views seem to imply that once a transgenic animal has been brought into existence it cannot be significantly harmed or wronged. It already lacks genetic integrity – a *telos* and/or naturalness – and so there is no possibility of its flourishing unharmed after the manner of its natural kind. Another implication of these views is that there is scarcely anything that can justify transgenic engineering, even if consequences of great value for humanity could be attained with little or no harm to the animals concerned. One of my purposes in this analysis is to investigate whether anything of value from these views survives these objections.

Another view which deserves early mention is that of Richard Ryder who maintains that the transgenic manipulation of animals is wrong only if it causes suffering to the animals involved (Ryder 1990). Ryder's 'sentientism' implies that causing suffering is normally wrong, so it is possible to harm and to wrong transgenic animals; and that is one of its strengths, and one of the weaknesses of the other kind of view. There is some limit to the amount of suffering that can justifiably be inflicted on such animals. But this view also has the weakness that no harm could be done to animals bred or engineered so as to be unable to feel pain or distress, whatever was done to them, and that no harm or wrong would be done in their original production either. So, if these are cases of possible harms or wrongs, there would seem to be reason either to favour the views of Fox (1990) or of Jonas (1984) after all; or at least, to supplement Ryder's (1990) view. Another purpose of this analysis is to investigate how Ryder's view should be supplemented or strengthened.

FUNDAMENTAL ISSUES ABOUT WRONGING AND HARM

It is sometimes held that bringing any creature into being can neither harm nor benefit it since a creature has to be there already before it can be harmed or benefited. If this were true, the second objection to Ryder's view (the objection that it *wrongly* makes unobjectionable the production of transgenic animals which lack feeling) would apparently be undetermined. But a few moments of reflection will suggest that it should rather be upheld, and will turn out to be time well spent in the matter of discovering ethical constraints applicable to transgenic manipulation.

The first reply to the claim that no creature can be benefited or harmed by bringing it into existence is that it is not plausibly true. As Derek Parfit (1984) has pointed out, we do not take this kind of view about death, much less about producing it; and theories which treat birth and death asymmetrically are themselves implausible. Again, it has traditionally been held to make sense to be thankful for one's existence, but this would make no sense if conferring existence were never a benefit. Further, if some qualities of life are worse than being dead, just as some are much better, then plausibly some are better and some are worse than non-existence, that is, than never having existed. There again, it is usually held that inflicting on someone a life of predictable misery is normally wrong, presumably because the subject of that life receives a life that is not worth living – or even a life that is worth not living. And if so, someone whose prospects at birth are of a life worth living or well worth living would seem to have been benefited by having been conceived and born. Thus harms and benefits are not restricted to events, activities and states of affairs arising *within* one's life. (Incidentally it should be noted that I have not assumed that inflicting a life of predictable misery on a creature is invariably wrong, in advance of considering possible justifications for doing so in cases of transgenic animals. What is here assumed is that the prospect of a life of predicable misery counts against generating that life, whether its subject is human or not; but this assumption seems unexceptionable.)

The second reply takes the form of a challenge to the assumption that actions can only be right or wrong because of benefits and/or harms to identifiable individuals. Once again it is Derek Parfit who has convincingly argued that this assumption would deprive most future-regarding actions of their rightness or wrongness, that actions with wide- or far-reaching impacts on future creatures have every bit as much of a moral status (whether positive or negative) as they were ever thought to, and that this status must thus depend on something other than the benefiting or harming of identifiable *individuals* (Parfit 1984, pp. 487–90). This is because the identities of future people and of future creatures of other species are often determined by current human action; as different individuals would come into being depending on what current agents do, there is no single individual who would exist whatever is now done, and who would thus stand to be benefited or harmed by the choice of one course of action rather than another. But the fact that no individual is benefited or harmed (or indeed wronged) does not absolve current agents of responsibility for the quality of life of whoever there will be, and for it being better or worse than it avoidably might have been. (Parfit's example of a policy of heavy present consumption of resources, to the detriment of their availability in the future, well illustrates a wrong without any identifiable victims.) Thus, even where no individual creature is harmed or wronged, there can still be actions that are wrong because of their bearing on future creatures.

SOME IMPLICATIONS

Now this excursion into ethical theory does not only show that bringing animals into being that are unable to undergo suffering is not above criticism. It has far-reaching implications concerning agents' responsibility for the lives which they bring into being and for their quality, some of which I hope to spell out later. It also gives the lie to the claim sometimes made that there is nothing wrong with producing animals for 'factory farming', or as experimental animals, or again as transgenic animals, *because they would not exist otherwise.* To this claim, two replies are now available. Where the animals concerned live lives of misery, lives which are not worth living, they are in any case harmed by having those lives inflicted on them; for these animals, non-existence would have been much better. Further, even if this is not accepted, it has become clear that current agents can act wrongly if they are responsible for lives being of low quality (or even of negative quality) when there could have been lives of higher quality instead, even when no individual is harmed or wronged. So the consideration that these animals would not otherwise have existed carries no weight whatsoever (any more than the consideration that slaves would not have existed otherwise justifies the breeding of slaves); and if their lives are predictably miserable, a very strong argument will be needed to justify their generation.

A further implication is that pain and suffering are not the only evils, for humans or for non-human animals. If they were, then the production of permanently anaesthetised transgenic animals would be beyond reproach. While it may not initially be easy to specify why such production would be wrong, it remains clear that the removal of the possibility of pain, far from making an affected creature's life worth living, would normally diminish its quality of life; for, as Darryl Macer (1989, pp. 233–4) has argued, the capacity to feel pain is normally a benefit. And, as the above excursion has shown, the avoidable reduction of quality of life is normally wrong.

Here it could be said, in defence of Ryder's (1990) view, that there is more to suffering than just pain; for suffering sometimes has emotional forms, themselves dependent on beliefs or on loss of sense of identity. Thus absence of suffering would involve a much more considerable psychological crippling than is at first apparent. While this should be granted, it is not sufficient to uphold 'sentientism'. First, there are kinds of harm that need not involve either physical or psychological suffering. As Alan Holland (1990) argues, this is possible for various kinds of injury, of disease or of incapacitation, and such harm is a proper object of concern for anyone prepared to take seriously the animal's point of view. Second, 'sentientism' is still open to the above criticisms, for it would now seem to imply that no wrong can be done to a transgenic animal incapable of both physical and psychological suffering, whereas we have just seen that such a creature can still undergo harm. It would also now seem to imply that there is never anything wrong with making a creature that is permanently anaesthetised in all these ways, whereas it has already been argued that, short of a very strong justification, that this would be entirely wrong.

It might now seem that appeal should be made to an imperative to respect nature or creatures' *telos*, in the manner of Jonas (1984) and Fox (1990). Is it not the evolutionary nature of each species which determines what counts as its flourishing, injury, disease, incapacitation or harm? While there is a lot in this last claim, evolutionary theory itself discloses dangers in regarding species' boundaries as fixed, and the same goes for the nature of species. As for respect, even if it were clear, as some Kantians suggest, that all evolutionary kinds should be respected, Holland (1990) has pointed out that Kantian respect for natural kinds does not preclude treating either fellow-humans or creatures instrumentally, as long as their ends are also respected. While it is manifest that transgenic manipulation essentially involves an instrumental attitude to animals, it does not invariably involve a neglect or subversion of what might be regarded as the implicit ends which as a result of evolution are embedded in their ways of life. There is also the problem mentioned at the start, that on these views, once the genetic integrity of species has been transgressed and transgenic animals have been produced, there would seem to be no way in which they can be either harmed or wronged, or in which wrong can be done in their regard. But this implication is unacceptable for now familiar reasons relating to suffering, injury and other kinds of harm, not to mention Parfitean reasons concerning quality of life.

QUALITY OF LIFE

Instead of the intuitive (and ultimately fruitless) appeal to nature, a better grasp of the ethics of transgenic manipulation can be attempted by further reflection on quality of life, along lines suggested by Holland (1990). This kind of reflection involves a willingness to move beyond an anthropocentric concern with the human interest in animals to readiness to take non-human interests seriously. This does not, however, involve attempting to enter into the subjectivity of animals, any more than a concern for other people involves telepathy. Nor does it involve imagining what it is like to be the creature in question. As has already been seen in connection with the difference between suffering and harm, there are many interests which do not turn on consciousness, and which are thus at least as manifest to an observer as to the creature in question.

Holland (1990), however, believes that these interests do not run out with harm and its avoidance. He imagines a strain of farm animals – pigs for example, bred to withstand the adverse conditions in which they, like their ancestors before them, are reared – having become more vegetable than animal. Such animals, holds Holland (1990) undergo a loss of freedom, defined as 'the capacity to exercise options'; and where the capacities of a species are significantly reduced, ethical objections are in place. There are two grounds for this, one of them stemming from human interests: natural life-forms function like old familiar friends, and thus serve the human need to preserve a sense of identity. But in any case, if we are prepared to take the view that life on earth constitutes a biotic community, respect for other members should be sufficient to rule out such reductions – Holland's (1990) second ground.

Holland's (1990) point about having an environment of 'old, familiar friends' is a valuable insight but could be satisfied by the preservation of enough members of old, familiar species, or of images of them, to maintain human psychological sanity at the same time as large numbers of semi-vegetable animals were being reared – out of sight in genetically sophisticated 'factory farms'. The central objection, then, stems from animal interests, although I doubt whether appeals to the biotic community strengthen either people's obligations or their readiness to acknowledge them. Nor is it clear that species as such warrant respect, as opposed to their members, valuable as it is that there should be members of every extant species, or virtually every such species.

There is also a further difficulty; the interests of the semi-vegetable animals are so slender, granted their impoverished genetic inheritance, that their treatment may easily neither hurt their feelings nor harm them nor blight their freedom. Holland (1990 pp. 171–2) appears to think otherwise, writing that: 'We ought to be very concerned by any developments which would diminish the general level of freedom of sentient animals'; but may not have meant the freedom for *individual* animals.

Could it be argued that such animals are wronged through having a semi-vegetable life inflicted on them? Yes, in theory, as has already been argued, but in practice the life of each of them might be immune enough from suffering to have a positive rather than a negative quality – a life, that is, which is just about worth living. Indeed in this regard they might be more fortunate than 'factory-farmed' pigs, whose lives it might be reasonable to regard as not worth living. And yet, though there is little or no loss to any individual, there is a loss. For where there might have been creatures with a higher or ampler level of capacities and the ability to exercise them, there is instead a lower or reduced level. And these are not just two abstract possibilities; in the absence of genetic manipulation, a different set of pigs with ampler capacities would actually have existed. Thus genetic manipulation would be responsible both for the identities and the natures of the semi-vegetable transgenic pigs. And this is wrong because, as has already been argued, it is wrong to cause there to be a lower and more restricted quality of life than would otherwise have been the case – even if no individual is worse off, and even if no single species features on both sides of the comparison. It is bad that there be crippled lives where there might have been restricted ones, even when being crippled is part of the synthesised nature of the cripples. And this remains true even if the less restricted lives might have been lives of privation led on a 'factory farm'.

The comparison is not basically one of freedom, and indeed the concept of alternative options open to an individual plays no part in the argument. It concerns rather the reduction of ranges of capacities, and the elimination of the more sophisticated capacities, including cognitive, connotive and affective ones. Sense can in this way be made of talk both of higher and of lower levels of capacities, and thus of deeds which involve their reduction. Accordingly, there is good reason to agree with Holland's (1990) conclusion that such a reduction is objectionable, even though the objection is based neither on sentience nor individual harm nor individual freedom.

It should here be remarked that actual transgenic animals are and will probably remain for the most part very different from the semi-vegetables of Holland's (1990) science fiction. For transgenic manipulation normally aims at what are, from the point of view of some human interest or other, improved versions of pigs or sheep, rather than semi-vegetable creatures far removed from pigs or sheep. And the interests and well-being of such animals remain in large part genetically determined by that of ordinary pigs or sheep. The more genetic changes are made the more unexpected side-effects appear, most of them disadvantageous to their bearers. Consequently there are strong ethical objections to practices that would approach anywhere near to those of Holland's (1990) imaginary scenario, objections in terms of the recognisable interests of the individual animals which would be produced long before such semi-vegetable products were reached.

GENETIC DIVERSITY

Holland (1990) deploys yet further arguments against transgenic manipulation concerning the loss of genetic diversity. This loss is likely to arise through cloning: 'where space which could have been occupied by a genetically distinct individual is occupied by an individual who is simply a genetic copy of another' (p. 173); and also because genes which are 'serviceable to humans' (p. 173) will be developed at the expense of life-forms which are not. His reasons why this could be bad may for the present purposes be set on one side; they concern the resilience which diversity confers on life-forms and ecosystems, and also the way in which reductions in diversity reverse the general direction of evolution to date, the trend which has made the biotic community possible. What is of greater relevance to my arguments here is Holland's (1990) objection to the standardised occupancy of 'life-space'. For if life-forms are eliminated whose distinctive capacities are unserviceable to humanity in favour of life-forms with serviceable capacities and those alone, that will frequently be open to the charge of the avoidable reduction of amplitude of capacities and the production of strains with a diminished range; and this already makes it unjustified unless there is some strong conflicting reason, like this being the only way to avert starvation among humans. While generalised protests in this connection such as the dangers of human arrogance have their place, they depend ultimately on arguments that seek to explain what precisely are the risks or the costs, such as the arguments adduced above. And those arguments are surely recognisable versions of the less focused positions of Fox (1990) and perhaps even of Jonas (1984) about the denaturing of creatures.

Holland's (1990) final argument concerns the accuracy of targeting of which genetic pest control is probably capable, and the dangers of removing 'entire threads from the fabric of the biotic community with unpredictable consequences' (p. 173). This is a weighty point; but as it turns largely on harms to existing creatures and the undermining of their species, it concerns a different aspect of the ethics of genetic engineering from those considered here. Perhaps, however, a remark illustrative of unpredictability and its significance may here be allowed. It is reliably reported that, because of nuclear experimentation during the 1950s and 1960s, the entire Agricultural Research Center at Beltsville, Maryland has been listed by the US Environmental Protection Agency (EPA) as potentially contaminated by radioactivity. It is thus possible that all the animals there, whether transgenic or 'controls', are affected by the unforeseen impacts of an earlier generation of experiments into the effects of fall-out on crops. Constant vigilance is clearly vital where such risks are a possibility.

Like Holland (1990) I do not believe that there is a conclusive argument for rejecting genetic engineering in principle. However, as I have argued here, there are a number of ethical arguments which require constraints, whether self-imposed by individual experimenters, research units and companies or imposed by legislation, as in the Environmental Protection Act of 1990. More particularly, I have attempted to sift the criticisms of animal genetic engineering of Fox (1990), Jonas (1984), Ryder (1990) and Holland (1990) and to show both (with Ryder and Holland) that transgenic animals can be harmed and wronged (against Ryder), that this applies even to insentient ones, and (with Fox and Jonas) that transgenic manipulation can be wrong even when no harm and no wrong befalls any resulting creature at all, not least (though not only) through reductions in the quality of life and in the range of capacities of whatever creatures we make or allow there to be: such reductions will sometimes outweigh any likely benefits.

18 I've Got You Under My Skin: Ethical Implications of the Practice of Xenografting

Dr Peter Wheale

XENOGRAFTING

In September 1993 a seven-organ transplant operation was carried out on 5-year old Laura Davies from Manchester by Professor Thomas Starzi, Director of the Pittsburgh Transplant Institute in the USA, in an attempt to save her life.

It was Professor Starzi, who, in 1963, carried out the first transplant of an animal organ to a human being when he put a baboon kidney into a man, and in 1992 he carried out the first transplant of an animal liver to a human. Professor Starzi believes that successful organ transplants from animals to humans heralds a new era in transplant surgery. 'Xenografting' is the transplantation of an organ or tissue between different species, and includes use of non-human organs and tissues for clinical transplantation into humans. (Human-to-human organ transplantation is called 'allografting'.)

The sad case of Laura Davies, who suffered from a condition known as 'shot-gut syndrome' which prevented her digesting food properly, highlights the enormous demand for transplant organs:

> There can be no doubt that a short-fall exists between the available supply of kidneys, livers, and hearts from cadaver sources and the number of persons awaiting transplants of these organs. Moreover, even if the current system of procuring organs from cadaver sources were modified so as to increase the efficiency of cadaver organ procurement, there would still exist a significant shortfall in the number of kidneys, livers and hearts available for transplantation to children and adults (Caplan 1985).

DNX, a US biotechnology company developing transgenic pigs, estimate that the international market in pigs' organs for xenografting is currently [1994] about US$ 6 billion annually.

The use of animals as a source of solid organs for the replacement or supplementation of the function of human hearts, livers and kidneys (and medical researchers are also giving consideration to other organs such as lungs, pancreas and bowel as candidate organs) is still a somewhat experimental procedure. Caplan (1985) suggests that our level of knowledge from research to date involving xenografts in human beings is approximately the same as that, for example, in the earliest stages in the testing of new cytotoxic drugs. Clinical studies conducted since the early 1960s indicate that the problems of xenograft rejection resemble those of allografting with respect to both cellular and humoral factors. There appears to be a strong correlation between the degree of genetic difference between the donor and recipient and the speed with which a xenograft is rejected.

These obstacles to xenografting do not dampen the enthusiasm of some molecular biologists for the future of xenografting using transgenic animals. Modern DNA technology (i.e. *micro*-genetic engineering) is able to contribute to xenografting in the production of drugs to inhibit rejection of donor organs and in procedures where tissues are grown in other animals for use in human beings. Scientists have been successful in genetically engineering a transgenic pig designed to reduce the likelihood of liver and heart transplant rejection in humans. For example, David White, a Cambridge molecular biologist applied for a patent on his transgenic pigs in 1989 and has since persuaded Sandoz, the Swiss pharmaceutical multinational, to support his transgenic work with pigs. DNX, the US biotechnology company referred to earlier, expects the first human xenograft of its genetically engineered pig's heart to be as early as 1996.

A recent development in the field of the central nervous system (CNS) tissue transplantation has suggested the use of human foetal CNS tissue from first-trimester abortions for xenografting. The benefit of such experiments is to expand our knowledge of the normal development mechanisms in the human CNS, and allow studies of various indices of maturation and CNS function. Efforts are also being directed toward understanding xenograft reactions in various other species, including from non-primate into primate (Reemtsma 1990).

Xenografting raises a number of ethical issues, including: whether or not a parent is competent to give informed consent on behalf of their child, where the child is to receive a non-human organ; whether or not there is adequate scientific knowledge to support such an experimental clinical procedure and the competency of medical staff to perform such operations; the adequacy or otherwise of regulations and research protocols controlling xenografting procedures; and the ethics of using non-human donors given the animal suffering and deaths which are involved.

I shall now consider this innovative but controversial area of health care by examining the moral status of the animals used in xenografting. In order to provide a framework for my analysis I shall avail myself of three very important theoretical approaches to moral investigation: rights-based morality; the goal-based approach of utilitarianism; and contractualism, which is often considered to be a duty-based system of values.

RIGHTS AND DUTIES

In medieval times maiming a person by surgically removing an organ constituted the common law offence of mayhem. Nowadays, providing consent is freely obtained from a fully informed donor, organ transplantation may be lawful (see, for example, Dworkin 1970). However, no individual is under a legal obligation to rescue another individual in this way. In *McFall* v. *Shimp*, a 1978 US case in which the plaintiff sought to request the court to order an injunction to compel the defendant to submit to tissue tests for the purpose of a bone marrow transplant, Judge Flaherty stated:

> For our law to *compel* the defendant to submit to an intrusion of his body would change every concept and principle upon which our society is founded. To do so would defeat the sanctity of the individual, and would impose a rule which would know no limit, and one could not imagine where the line would be drawn.

We also know that the saving of one's own life, even in extreme circumstances, at the expense of the lives of other individuals is morally wrong. In *Regina* v. *Dudley and Stephens*, an 1884 case in which two starving men adrift at sea had eaten a cabin boy in order to survive, the defendants were found guilty of murder (discussed in Frey 1983). If this prohibition on the involuntary use of another's organs is the position with regard to human beings, in the course of the practice of xenografting, where the donor is an animal, from a moral perspective the question must arise as to whether it is ethical to kill an animal to save a human life.

Most people have accepted that species is the morally relevant criterion to determine membership of the moral community, and the dominant intuition is proclaimed to be: that whilst animals may have some moral significance, their significance is always less than that of humans, and we are, therefore, always entitled to prefer our interests to theirs (see Cartwright 1991). However, in *The Case for Animal Rights*, Tom Regan (1983) argues that all animals have intrinsic worth and are thus worthy of respect. He asserts that at least some non-human animals do have inviolable moral rights, the most important of which is the right to life.

Regan (1983) argues that in failing to extend rights to non-human animals we are engaging in species discrimination. The idea here is that if some animals share similar characteristics with humans (and studies in neurobiology have furnished evidence that humans share many cognitive and perceptual attributes with other members of the animal kingdom – see, for example, Griffin 1992) then why should they not be granted the same rights we grant to humans? Human beings are said to maintain a discriminatory policy towards non-human life – speciesism – a prejudice analogous with racial or sexual discrimination, which some claim (Richard Ryder 1990 for example) insulates animal experimenters from the ethical consequences of their actions by reinforcing the perceived divide between us and other animals.

Joseph Raz has defined a right as attributable to an individual if an aspect of his well-being or interests is a sufficient reason for holding some other person to be under a duty (Raz 1986). Sentience is sometimes argued as a criterion for moral status.

> If a being can feel pain, then it constitutes a center of consciousness to which justification can be addressed. Feeling pain is a clear way in which the being can be worse off;

> having its pain alleviated is a way in which it can be benefited; and these are forms of weal and woe which seem directly comparable to our own (Scanlon 1985).

If we accept that (at least certain) non-human animals have a right to life – a right to autonomy – then, on the Razian principle we must accept that we have a duty not to interfere with the animal's exercising of its right to life, except perhaps in emergency situations. Of course, many rights-based moralists would argue that we can protect an individual's interests without according him rights. Thus, if we accept this view, protecting an individual's *interests* will not always give rise to rights or therefore to corresponding duties.

UTILITARIANISM

Utilitarianism, as a moral philosophy, determines that which is deemed morally right according to that which maximises happiness (or utility) for the greatest number of individuals. 'Philosophical utilitarianism' is thus a particular philosophical thesis about the subject matter of morality in which the only fundamental moral facts are facts about individual well-being (Scanlon 1985). In its traditional and purest form, utilitarianism determines what actions are morally right by examining and weighing their respective consequences. According to Jeremy Bentham, the principle of utility is that principle which approves or disapproves of every action whatsoever according to the tendency which it appears to have to augment or diminish the aggregate happiness of the community as a whole. Actions are right in proportion as they produce 'good', and the act-utilitarian is prepared to break rules and to violate rights in order to maximise 'good' (or to minimise 'misery') – variously considered to be happiness, pleasure, the sum of people's preferences, welfare or utility.

In directly judging individual actions, the utilitarian, or more strictly speaking, the *act*-utilitarian, simply accepts the consequences of his principles, and is prepared to dismiss any contrary 'intuitions'. Whereas the moralist concerned with rights and duties would appraise an action *intrinsically* in terms of its moral character, the utilitarian appraises it *extrinsically*, that is, in terms of its consequences.

In his book *Animal Liberation* Peter Singer (1977) argues that it is arbitrary to restrict the principle of equal consideration of interests to our own species and that the moral domain should extend to all those whose welfare can be affected by our actions. Accordingly, all sentient creatures should be included as we can improve or demise their welfare (see also Rollin 1981).

There is, however, a marked distinction between having regard for the welfare of animals and believing that animals have ('natural') inviolable moral rights. As a utilitarian, Peter Singer (1977) prefers to avoid referring to the moral *rights* of animals, at least in so far as these are construed as claims which we may sometimes override on purely utilitarian considerations. According to this moral philosophy the 'costs' in terms of animal suffering must be weighed or balanced against the 'benefits' to humans. One problem in attempting such an analysis is that, given that our pluralist societies are highly diverse, each individual's utility will be unique, and even if there were agreement on the predicted consequences of an action it is unlikely that consensus could be reached on the valuations required to weigh the costs and benefits of the consequences.

The key factor in this balancing approach is that the interests of animals are being given equal consideration with those of humans. There will, therefore, be circumstances when animal suffering can be justified, but also circumstances could arise, at least in theory, where we would allow human suffering for the welfare of animals. It is because the interests of humans and non-humans are given equal consideration that weighing of the costs of suffering and the benefits to be derived from that avoidable suffering are not predetermined. Accordingly, as Cartwright (1991) argues, whilst utilitarianism would appear to support the first component of the dominant intuition referred to above, that is, that non-human animals matter morally, it does not support the latter component of the dominant intuition, that is, that they *always* matter less than human beings.

Whilst this outcome may be palatable to some individuals whose primary concern is animal welfare, there are other fundamental problems with utilitarianism as our guiding moral philosophy.

First, although in theory human suffering is balanced

against animal suffering, in practice, existing laws strike a balance dramatically in favour of human interests and accept that animals may be used in experiments as long as there is *some* benefit expected from their use. With regard to non-human animals, some argue that, as long as those who do the balancing regard virtually *any* benefit as adequate justification of animal suffering, there will be no *effective* regulatory constraint on animal experimentation (see, for example, Francione 1990).

Secondly, because maximising aggregate happiness may be at the expense of others, a utilitarian analysis often produces results which run counter to our dominant intuitions. For example, suppose it were to appear that, on balance, slavery maximises social utility, then, in principle at least, the traditional utilitarian would allow it as morally right. If the master was believed to be of greater social value than the slave, we could imagine circumstances when a utilitarian would advise the use of the slave as a source of organs for transplantation to save the life of the master! Thus, utilitarianism may require us to do or condone things we could find psychologically or emotionally unacceptable – acts which violate the principle of 'psychological realism' according to Griffin (1992).

The sole objective of utilitarian morality is to maximise the aggregate 'good' if needs be at the expense of the individual, thus violating the principle of the separability of persons. For example, it is unacceptable to most of us that circumstances could arise where it would be considered that our loved ones should be sacrificed for the general good. It is for these reasons that L. W. Sumner (1990 Chapter 6) suggests that the utilitarian approach generally misrepresents our nature, reducing us to creatures concerned primarily with desire-satisfaction, and therefore *understating* the value we place on individual freedom and equality.

In an attempt to address the sorts of criticisms described earlier, some utilitarians argue that it is not the consequences of actions that are being evaluated but the moral rules or guidelines themselves. In this way, these *indirect* or *rule* utilitarians argue that the welfare of the individual can be protected from the aggregate good of society. John Stuart Mill's 'humanitarian' utilitarianism, for example, equates maximising happiness as maximising the sum total of each individual's preferences, providing their actions do not harm others. This form of rule-utilitarianism embodies a respect for the inviolability of *individual* autonomy.

Suppose rule-utilitarians suggest, for example, that the following *rule* be imposed: 'that humans are always to be preferred to other animals'. This refinement would then constrain the hypothetical balancing of the 'costs' and 'benefits' of the predicted consequences of animal experimentation in whatever circumstances to produce moral outcomes always favourable to the welfare of human beings. On close examination, however, we see that such a rule-utilitarianism merely creates a two-tier moral system. On one level it attempts rather to protect the individual from emotionally and psychologically unacceptable outcomes, and yet retains on a 'higher' level of moral theorising the proposition that the general happiness, which constitutes the ultimate moral goal of utilitarianism, is said to compensate for the seemingly unjust misery, or in some circumstances even death, of other individuals (Mackie 1984). I contend that anyone able to sustain such a 'schizophrenic' moral position is suffering from a state of mind which the psychologist, Eric Fromm referred to in another context as the 'pathology of normalcy' (Fromm 1970).

This brief analysis of utilitarianism (both act-utilitarianism and rule-utilitarianism) has revealed the approach to be of limited use because it is not only liable to produce inconsistent moral positions but can be impracticable as a means of morally justifying many (hypothetical) actions. As a method of justifying the infliction of suffering on a particular group or individual in order to benefit another group or individual I consider we are therefore ultimately compelled to reject utilitarianism as a reliable moral framework.

CONTRACTUALISM

Contractualist theories attempt to ground principles of morality in some hypothetical procedure of collective choice. Of course, within this framework it may be, as Scanlon (1985) points out, that *some* moral questions could be properly settled by appeal to the utilitarian principle of maximum aggregate well-being, even though for contractualists this is not the sole or ultimate standard of justification. For the contractualist moral wrongness may be stated as follows:

> An act is wrong if its performance under the circumstances would be disallowed by any system of rules for the general regulation of behaviour which no one could reasonably reject as a basis for informed, unforced general agreement. (Scanlon 1985).

In *A Theory of Justice* John Rawls (1972) gives an exposition of a contractualist theory of morals. Rawls's idea is that, differences of opinion and status lead individuals to make conflicting claims on one another and that therefore:

> A set of principles is required to judge between social arrangements that shape this division of advantages. Thus the role of the principles of justice is to assign rights and duties in the basic structure of society and to specify the manner in which institutions are to influence the overall distribution of the returns from social cooperation. (Rawls 1985).

Rawls (1972) suggests that the method by which a moral theory becomes acceptable is, what he calls, 'reflective equilibrium'. Rawls's idea is that any theory of morals must accommodate the community's dominant moral intuitions (or be itself so persuasive that upon reflection the community is moved to abandon those intuitions which are incompatible with the proposed theory of morals). Of course, in any such notion of equilibrium, moral theory and dominant intuitions must also be socially *reflexive* and dynamic, rather than static, in order that cultural change be accommodated and morality evolve over time.

Rawls (1972) asks us to imagine that we were founding a new society and invites us to consider what set of moral rules we would rationally prefer above all others. As parties to this hypothetical contract we are to agree upon a set of reciprocal obligations which we find it in our mutual self-interest to adopt as a binding set of moral rules which are to guide all subsequent social behaviour. Importantly, in order to rule out partiality, all who are party to this hypothetical contract must choose their moral system under conditions of uncertainty (under a 'veil of ignorance') as to what their role, identity and status in society will be.

> First of all no-one knows his place in society, his class or position or social status; nor does he know his fortune in the distribution of natural assets and abilities, his intelligence and strength, and the like (Rawls 1972 p. 136).

Rawls calls these circumstances the 'original position' and it is supposed to enable individuals to choose the most just and feasible morality which properly balances their respective competing claims and interests (see, for example, Kukathas and Pettit 1990). The 'veil of ignorance', Rawls (1972 p. 140) argues, will ensure that all rational individuals will unanimously agree upon a fair and feasible conception of justice. It must be said that Rawls is imposing certain conditions on the natures of the rational parties to the contract: he assumes that they must not be influenced by envy, and by implication, for example, that they have a sense of fairness. (For a critical analysis of Rawls's assumptions see Dworkin 1977 Chapter 6).

From this 'original position', Rawls argues that the rational parties to the social contract would all agree to be governed by two principles of justice: the first guaranteeing fundamental individual liberties to everyone; and the second ensuring socio-economic inequalities are arranged so as to offer the greatest possible benefit to the worst off in society, while upholding conditions of fair equality of opportunity to all (Rawls 1972 p. 302). The adoption of these two principles of justice gives public support to the self-respect of individual members of society and is therefore a Kantian conception of equality (Rawls 1985) in the sense that people must be treated as ends and 'not merely as means' to the greater collective good.

Economic fairness includes requiring one generation to save for the welfare of future generations. (A requirement which happens, incidentally, to be compatible with the principle of 'sustainability' (see Brundtland 1987). Sustainable development, with its corollary of sustainable income (see, for example, David Pearce 1991), requires that each generation pass on to the next generation an undiminished aggregate of capital assets, including certain environmental assets such as our stock of biological diversity, which should be inviolable.)

Who under Rawls's contractualist theory qualifies as a

rational party? If we accept that every right we assert, regardless of its content, imposes some duty or disability on those against whom it is held (i.e. contractual reciprocity), then it follows that in order to qualify as a rational party the individual must be capable of exercising normative powers, that is, must possess Lockean (after the philosopher John Locke) 'self consciousness'. It would appear that under this notion of contractualism only rational agents can be objects of moral rights (see Sumner 1990 p. 202). Moreover, if agency (or autonomy) is a precondition of having rights in Rawls's contractual choice model, then logically this precludes rights to babies, foetuses, and severely mentally handicapped persons, and certainly to all non-human life (Rawls 1972 p. 512). However, this outcome is incompatible with Rawls's own theory of 'reflective equilibrium' (referred to earlier) because such a position cannot accommodate the dominant intuition which attributes a degree of moral status to individuals who are not autonomous and are therefore incapable of reciprocal contractual relationships.

Rawls's argument seems to have cornered us into the unsatisfactory situation whereby we can only either accept his position and reject certain of our moral intuitions, or, alternatively, retain them and dispense with his theory. However, before rejecting Rawls's contractualist theory I wish to re-examine the status of the rational agent reciprocity requirements.

In the medico-legal context, in cases where a patient is unable to give consent to an emergency medical procedure, the 'substituted judgement' doctrine is sometimes applied (it is particularly prevalent in the USA). This doctrine requires the court to 'don the mantle of the incompetent' and to 'substitute itself as nearly as may be for the incompetent, and to act upon the motives and considerations as would have moved her' (Robertson 1976). It appears to me that Rawls takes up a similar position to that of the 'substituted judgement' doctrine by conceding that, in order to maintain the integrity of incompetent persons, we should act towards them 'as we have reason to believe they would choose for themselves if they were [capable] of reason and deciding rationally' (Rawls 1972 p. 209). Thus, Rawls circumvents being cornered by his reciprocity requirement by relying on a form of 'substituted judgement' approach to the social contract.

If the rational parties to the 'social contract' are obliged to use 'substituted judgement' to imagine alternative unique 'qualities of life', then it seems to me perfectly logical for the rational parties in the 'original position' to consider the interests which they would have if, when the 'veil of ignorance' is removed, they were to find themselves not only in the position of foetuses or severely mentally handicapped individuals, but in the position of (at least certain) non-human animals. In this way, and contrary to Rawls's viewpoint, I believe, we *can* logically extend the Rawlsian contractual choice model to address the interests of at least some forms of 'non-rational' life.

In order to illustrate the plausibility of this argument, take for example the fact that human beings are mammals, and as such, in a Darwinian sense, are biologically closely related to all other mammals. Using whatever scientific knowledge about mammals that exists (and scientists in recent years have demonstrated that it is no longer possible to deny the existence of forms of thought, and even the capacity for cultural transmission, in many living species), it surely seems at least as feasible for rational individuals to reach agreement on what the basic interests of different species of mammal are thought to be, and therefore, by inference, what their basic rights should be, as it is to determine such matters for non-competent humans.

Even if we agree that a theory of rights which adopts the choice model can make no sense of the rights of non-human animals or foetuses, or infants or young children or the severely mentally handicapped, it can accomplish essentially the same objective by making them the beneficiaries of our protective duties (see Sumner 1990 p. 204). Therefore, notwithstanding the 'agency' precondition, it would seem that even under Rawls's contractual model it *is* logically possible for us to choose to have relational duties to any creature capable of being harmed or benefited, and, therefore, for any such creature to have claims against us. Indeed, in practice, rights are frequently asserted and defended by ideologically motivated groups which possess strength (that is, by advocates) on behalf of other parties who lack the strength to assert such rights on their own account (see, for example, Pocar 1990). Agents in the 'original position' would at least be able to imagine what it would be like to be an animal rights campaigner for example and therefore to wish to accede to some state protection of non-human animal interests.

What appears to be culturally repugnant is not the production of suffering as such, but any suffering not supported by a justification that is culturally sufficient according to the accepted system of values. It is these values that utilitarians can so often find themselves unwittingly supporting in defence of inflicting suffering at the expense of one group or individual for the 'greater' good of another group or individual considering that all values and standards belong within human thinking and practice and have no independent objective existence. Philosophers and sociologists of science have argued persuasively that factual assertions are completely contingent on theoretical assumptions, and that observation itself is always subject to interpretation (for an accessible discussion of this argument see Boyle *et al.* 1986 Chapter 3). From this perspective our values are most certainly socially constructed.

Our values and standards are stabilised as (normative) expectations through repeated communications (see Luhmann 1986). This is often done by socially powerful groups:

> Witness the ease with which it proved possible to deny any 'reason' to women and to Africans sold as slaves. Domestic and laboratory animals are considered as property in the same way that slaves were formerly treated (Pocar 1990).

Our justification, then, for the second component of our dominant intuition concerning the moral status of non-humans – that is, that the moral significance of non-human life is always less than that of humans – is, in reality, merely an ideological construct. In this instance an ideology in which the animal is conceived of as a commodity, and its body parts therefore as property to be disposed of in any way we choose.

I am not, however, suggesting that because rights and moral standards are not grounded in 'nature' and are not therefore discoverable by reason, that we are precluded from asserting that certain rights *should* be adopted as fundamental and universal moral principles (Mackie 1984). On the contrary, once we accept the idea that at least some non-human animals could *effectively* have moral rights, then our instrumental attitude to non-human life must radically change. For example, if we accept that a baboon is an individual with a right to exist not contingent on his or her providing a benefit to humans, then it is far less difficult to see why the death of baboons used in xenografting, for example, is morally unjustifiable (Francione 1990).

Certain social developments are providing evidence that public values and attitudes to non-human life *are* changing. I contend that a process of 'reflexive dialectics' between public pressure groups, including animal welfare campaigners and environmentalists, is bringing about a new 'reflective equilibrium' between our dominant intuitions and morality. In recent times this social process has resulted in the expansion of the scope of bioethics to include animal welfare and environmental ethics, as well as medical ethics (see, for example, STOA 1992; Wheale and McNally 1994). This particular development is an example of the effects of the successful communication of arguments in favour of environmental protection and animal rights and provides us with evidence of the changing social attitudes towards non-human life and the environment. Such attitudes become a source of moral motivation capable of substantiating one's actions on grounds that others cannot reasonably reject.

PUBLIC OPINION

New innovations in science and technology often test the 'robustness' of our traditional moral values and ethical positions. Xenografting, defined as cross-species transplantation, is such an innovation and has raised a number of controversial ethical questions not least of which is the question of the morality of killing animals in an attempt to save the lives of human beings. It is the issues relating to this question that I have attempted to address here.

Many animal welfare campaigners claim that animals have intrinsic and inviolable rights. I have attempted to explore whether or not this claim can be logically sustained within utilitarian and/or contractualist moral theory. My analysis has revealed that animals are accorded some moral status within a utilitarian framework but I have argued that we should reject utilitarianism as a general ethical justification for our actions because it can direct us to assert moral justifications for actions which are, to most of us at least, morally, psychologically and emotionally unacceptable. In applying Rawls's principles of justice as outlined in his book *A Theory of Justice* (Rawls 1972) I have reasoned (contrary to Rawls) that it *is* logically possible within such a contractualist framework for the rational parties (in the 'original position') in choosing the most just system of moral priciples to choose to have relational duties to any creature capable of being harmed or benefited, and for any such creature to have claims against us and that, in this way, non-human animals *can*, in effect, be granted rights.

I have argued that, just as it is logical to extend non-reciprocal rights to incompetent humans, it is equally logical to extend similar rights to, at least some non-human animals, for example, certain mammals. Furthermore, that even were this particular argument for animal rights rejected, that it is still logical within the Rawlsian scheme to allocate such rights because some of the rational agents in the original position will become animal rights campaigners and therefore require certain rights to be conceded to animals as part of their social contractual arrangements with their fellow humans.

I have also suggested that the continuing justification for our instrumental relationship to non-human life is, in reality, an ideological construct – in this instance that non-human life is a commodity to be treated as mere property – but that, however, as a social construct it is therefore susceptible to change. Such changes over time are consistent with the notion of a 'reflective equilibrium' which Rawls suggests must obtain between a society's moral theory and its dominant intuitions because such an equilibrium must needs be dynamic and *reflexive* in order to accommodate inevitable cultural changes.

Finally, I have asserted that a change in public values and attitudes to non-human life is currently discernable and that this is evidenced by the recent expansion of the scope of bioethics to include animal and environmental ethics as well as medical ethics. Western governments, in as far as they have been prepared to adopt this more broadly defined bioethics to evaluate new technologies are surely displaying their increased willingness to concede an expansion of rights to at least some forms of non-human life.

Acknowledgements

I wish to thank Ruth McNally for her most helpful comments on an earlier draft of this paper. I alone, of course, accept full responsibility for the views expressed here.

19 Public Policy Issues Relating to Animal Biotechnology

Dr Maurice Lex

PUBLIC POLICY

This analysis of public policy issues relating to animal biotechnology in the European Union (EU) begins by raising several questions. What role should a public authority such as the European Commission play in response to what, in essence, is a breakthrough or novel discovery in science? To what extent should society seek to control or regulate the very innovations it has earlier sought to create or to catalyse? To what extent should socioeconomic and ethical considerations be taken into account in the assessment of publicly funded research and its subsequent commercialisation? This contribution will not attempt to provide all the answers but to serve in some small way to provoke discussion and debate in an open and democratic society.

Human beings have long sought to interfere with the genetics of animals; to improve them to suit human purposes; to make them more suitable for human service. No doubt over the years some disastrous selections or crosses have been made, many of which may have caused animal suffering but most of which would have been lethal. On the whole, selective breeding of animals, particularly of farm animals, has resulted in many changes to what 'mother nature' originally designed and most of them have led to 'improvements' without causing excessive animal suffering.

In 1980 there was a breakthrough in science that enables us to cross animal species barriers and insert genes from almost any organism into animals. Since then, the humble mouse has been the preferred model to demonstrate the possibility of transferring and expressing specific genes. The technique has now been extended to many other animal species including the 'traditional' farm animals.

Transgenic animals are made for a number of reasons including fundamental research, biomedical research, clinical transplantation and production. For fundamental research, transgenesis may be undertaken as research *per se*, or for the construction of specific animal models to study fundamental biological phenomena. For biomedical research transgenesis is used to construct specific animal models to study human diseases, for instance cancer, and specific animal models for the testing of pharmaceuticals, chemicals and cosmetics. For clinical transplantation, transgenesis is used for the development of animals with 'humanised' organs (heart, liver, kidneys). And for production, transgenesis is used to obtain human therapeutics such as Factor VIII and haemoglobin, for increased or improved meat production, and for improved animal health and disease resistance.

Innovation in research has rapidly led to commercialisation, and to the establishment of a European Trade Association for Advanced Animal Breeding (ETAAAB) involving some 13 members in 1992.

How should a public authority such as the European Commission respond to such developments? This has been annunciated in the Commission's April 1991 Communication to the Parliament and the Council entitled 'Promoting the competitive environment for the industrial activities based on biotechnology within the Community' (SEC(91)629). The actions and developments since then have been described in a Biotechnology Coordination Committee report 'Biotechnology after the 1991 Communication: A Stock Taking' (BCC 1992) and in the Commission's White Paper (1993).

The main public policy issues raised by biotechnology in Europe relate to the following topics: research, development and innovation; ethical considerations; the legal and regulatory framework; standards; and intellectual property.

RESEARCH, DEVELOPMENT AND INNOVATION

The European Commission has supported research for biotechnology under the 'Biotechnology Research for Innovation, Development and Growth in Europe' (BRIDGE) 1990–93. There are two types of BRIDGE research projects: N projects and T projects. N projects tend to be smaller, stand-alone projects while T projects are larger, more targeted projects, coordinated through an administrative mechanism and collaboration between several laboratories. Both N and T projects are concerned with research related to animals and to animal cells.

A considerable number of projects come under the heading: 'Prenormative research: *in vitro* evaluation of the toxicity and pharmacological activity of molecules' and 3.9 million ECU (MECU) has been allocated over a 3-year period to support eight N projects in this area.

A major T project relates to Animal Cell Biotechnology. It aims to improve the processes currently employed to produce products of major therapeutic value, such as monoclonal antibodies, or tissue plasminogen activator (TPA) (for treatment of heart attacks) in animal cell cultures. A total of 2.5 MECU over 2 years is committed to 6 projects. An Animal Cell Technology Industrial Platform (ACTIP) composed of over 20 firms active in this area has been set up to interact with the T project researchers and aid in technology transfer.

Support from the European Commission is also provided under the Biomedical and Health Research Programme of the Third Research and Technological Development (RTD) Framework Programme (FP3) (1990–94). The Biomedical and Health Research Programme has a 'Concerted Action' project involving 16 laboratories throughout Europe. In 1992, this project organised a meeting in Edinburgh to discuss priorities for transgenic animal research as Models for Human Diseases.

A further source of European Commission support for animal biotechnology is the FP3's Biotechnology Programme 1992–94, which funds such research under the following programme lines: cellular regeneration, reproduction and development of living organisms; and metabolism of animals and essential physiological traits.*

In May 1992, a European Centre for the Validation of Alter-

---

* Editors' update on European Commission support for biotechnology RTD: Biotechnological research is a growth area for EC RTD funding. Funding rose from 120 MECU under FP2 to 167 MECU under FP3. The budget envelope for Life Sciences and Technologies under FP4 (1994–98) is 1080 MECU, of which almost 50 per cent will be spent on biotechnological research under the following programme lines: biotechnology; biomedicine and health; and agriculture and fisheries.

native Methods to Animal Testing (ECVAM) was established at the European Community's (EC's) Joint Research Centre site in Ispra, Italy. As a unit of the Environmental Institute, it seeks to tackle the difficult progression from the development of an alternative test to its acceptance as part of the regulatory testing system. There are proposals that it should set up a central database on Europe-wide experimentation and results.

ETHICAL CONSIDERATIONS

In April 1991, the three Presidents (of Parliament, Council and Commission) agreed that all of the FP3's RTD programmes should incorporate socioeconomic assessment and assessment of technological risks. In three programmes – Environment, Biomedical and Health research (which includes human genome research), and Biotechnology – ethical assessment is also explicitly included.

In the case of the Biotechnology Programme, at the insistent demand of the European Parliament in approving the Programme, in addition to the laboratory research projects mentioned earlier, up to 3 per cent of the Programme's total budget (167 MECU) will be devoted to 'horizontal activities' – those which are common to all areas of the Programme – to assess the ethical and socioeconomic effects and technological risks from biotechnology. It is likely that one or more of these horizontal activities, which are implemented by means of e.g. study contracts and 'concerted actions', will address the issue of transgenic animals.

The ethical, social, legal and economic implications of biotechnology will be monitored and studied with a view to contributing to the clarification of issues such as the diversity of perceptions on benefits and possible risks of biotechnology and the factors influencing the acceptability of scientific results. A further area of interest will be the role of public policies, including research policies, in influencing these issues. The potential for using the applications of biotechnology research and development to enhance social progress and economic development through innovation in agriculture, medicine and industry will also be studied. A multidisciplinary approach involving representatives from the various sciences, professions and activities concerned will be encouraged.

The Commission recognises that biotechnology is raising ethical considerations which are attracting considerable public interest and debate, some of it confused. Such confusion can adversely influence the whole climate for the development of the technology by discouraging students from choosing biotechnology research projects and providing a disincentive for industrial investment. In its April 1991 Communication (SEC(91)629) the Commission sought to differentiate between different categories of issues. One approach has been to set up a Group of Advisers on the Ethical Implications of Biotechnology.

The Group's terms of references are as follows: identification and definition of ethical issues raised by biotechnology; appraisal of the ethical aspects of EC activities in the field of biotechnology and their potential impact on society and the individual; advising the Commission as regards the ethical aspects of biotechnology with a view to improving public understanding and acceptance of it. In performing its tasks the Group provides the Commission with appraisals of the potential ethical impact of activities based on biotechnology. The Commission may request the Group's opinion on a particular issue. The Group is also able to submit reports to the Commission on its own initiative and present its opinions on all general matters of an ethical nature.

The Group's work commenced in March 1992, with the first task of appraising the ethical impact of the application of growth enhancers in agriculture and fisheries. The Commission has subsequently submitted a request to the Group for their opinion on ethics and biotechnology for animals. The kind of ethical issues raised include whether limits should be imposed on gene transfer experiments (for example experiments in transferring human genes into an animal) in relation to the integrity of animals in terms of health and welfare, the maintenance of the identity of the species, and its ability to reproduce. To what extent should animals be subordinated to human needs bearing in mind the difficulty in assessing the relative value of the advantages to humans weighed against the possible discomfort or suffering to animals?

In this field it is seen as particularly important for the Commission to maintain close contact with the substantial and continuing work of the Council of Europe (in the case of animal welfare conventions), national bioethical committees such as the UK's Nuffield Council on Bioethics and the MAFF's Farm Animal Welfare Council (FAWC) and the European Parliament's Scientific and Technological Options Assessment (STOA) Unit's study on bioethics (STOA 1992) which has given specific consideration to transgenic animals.*

THE LEGAL AND REGULATORY FRAMEWORK

EC legislation addresses animal welfare in the contexts of experimental and other scientific purposes, farming, the production of transgenic animals, the protection of human health and the environment, research and development, and marketing. The EC has worked very closely with the Council of Europe on the subject of animal welfare and is a contracting party to several Council of Europe Conventions.

For transgenic animals that are considered to be kept in contained use, no EC legislation exists. Animals used for experimental purposes are covered by EC Directive 86/609/EEC. The objective of this Directive is to 'harmonise' – to eliminate disparities between – national laws for the protection of animals used for experimental or scientific purposes. This Directive was adopted in November 1986 and came into force in November 1989. Essentially, it provides a framework for the control of the use of animals in research, such that premises, personnel and procedures may be subject to prior authorisation before experiments can take place.

The scope of Directive 86/609/EEC is defined in Article 3 where it is stated that experiments on animals may only be

* Editors' update on the Group of Advisors on the Ethical Implications of Biotechnology: By 1994 the Group had adopted opinions on the ethical implications of the use of performance-enhancers in agriculture and fisheries; products derived from human blood or human plasma; and ethical questions arising from the Commission's proposal for a Council Directive on the legal protection of biotechnological inventions. Work was in progress on the topics of transgenic animals and gene therapy. The initial membership of 6 individuals was expanded in 1994 to 9, and its mandate was altered to comply with the recommendations of the EU's White Paper on growth, competitiveness and employment (European Commission 1993 p. 118).

carried out for the purposes of:

(a) development, manufacture, quality effectiveness and safety testing of drugs, foodstuffs and other substances or products; and
(b) for the protection of the natural environment.

The Directive's general provisions are as follows: endangered species may only be used in exceptional circumstances (Article 4); Guidelines for Accommodation and Care of Animals are given in Article 5 and standards for care and accommodation are exemplified in Annex II; Article 19 is a special provision for the protection of cats, dogs and primates; and Article 22 states that unnecessary duplication of experiments should be avoided wherever possible.

The principles of the Directive are modelled closely on those of the Council of Europe Convention. The framework ensures that the number of animals used for experimental purposes is reduced to a minimum, that the animals are properly housed and cared for, and that no pain, suffering, distress or lasting harm are inflicted unnecessarily. Where pain, suffering or distress are unavoidable, they should be kept to a minimum. Similar provisions are applied to breeding and supply of laboratory animals, and special controls exist for experiments involving anaesthesia and for certain species deemed to require special protection.

A decision of the European Council in 1978 established the EC as a contracting party to the 'European Convention for the Protection of Animals Kept for Farming Purposes'. In 1991 the Convention was amended by a 'Protocol of Amendment' to take account of biotechnology and on-farm killing. The amended Article 1 now reads as follows:

> This Convention shall apply to the breeding, keeping, care and housing of animals and in particular to animals in intensive stock-farming systems. For the purposes of this Convention 'animals' shall mean animals bred or kept for the production of food, wool, skin or fur, or for other farming purposes, including animals produced as a result of genetic modifications or novel genetic combinations. 'Intensive stock farming' systems shall mean husbandry methods in which animals are kept in such numbers or density, or at

> such production levels that their health and welfare depend upon frequent human attention.

A new Article 3 now inserted into the Convention reads as follows:

> Natural or artificial breeding or breeding procedures which cause or are likely to cause suffering or injury to any of the animals involved shall not be practised; no animal shall be kept for farming purposes unless it can be reasonably expected, on the basis of its phenotype or genotype, that it can be kept without detrimental effects on its health or welfare.

Council Directive 90/220/EEC on the deliberate release into the environment of genetically modified organisms, adopted in April 1990, came into force on 23 October 1991. This Directive requires an environmental risk assessment, notification and permit for all transgenic animals to be released into the environment both for research and development purposes and for placing on the market. Each Member State must set up a Competent Authority to implement the Directive.

In addition to the above, there are a number of draft Directives or Regulations relating to transgenic animals. For example, Directorate-General VI (DGVI) (Agriculture) has prepared a draft Proposal for a Council Regulation on the Placing on the Market of Genetically Modified Farm Animals and Semen, Ova, and Embryos. Where the transgenic animals are food animals, there exists the possibility that novel elements could be brought into the food supply, in which case they could be covered under the draft proposal for a European Parliament and Council Regulation on novel foods and novel food ingredients (COM(93)631). Where genetically modified animals or products are to be placed on the market, both of the above drafts would require an environmental risk assessment similar to that required under the deliberate release Directive (90/220/EEC) as well as a copy of the written consent required under Directive 90/220/EEC from the Competent Authority to the deliberate release of the transgenic animal for research and development purposes. Sectoral (product-based or 'vertical') legislation, such as provided for in the above drafts, fulfils the Commission's commitment to one integrated assessment and notification procedure to cover all that is required for product authorisation, the so-called, 'one-door/one-key' policy.

STANDARDS

The Commission Communication of April 1991 (SEC(91)629) gave a commitment to European standardisation not only to help create a standard technical environment for the further development of biotechnology but to complement and fulfil the regulatory framework. In its subsequent 'stock-taking' note (BCC 1992) the Biotechnology Coordination Committee (BCC) reports that a clear and precise mandate has been given to the European Committee for Standardisation (CEN: Comité Européen de Normalisation). The question is whether there is a role for 'standards' in the conduct and commercialisation of transgenic animals research. It would appear that a very good case can be made for a 'Code of Good Practice', or 'Performance Standards for Research'. Both industrial and academic researchers must ensure that transgenesis does not result in unnecessary animal suffering, pain or disease; in the creation of 'unacceptable' hybrid organisms with abnormal behaviour; or in unforeseen impacts on the environment. The development of such technical standards to complement the existing legislation on animal protection, both for experimental and farming purposes, would serve to reassure a sceptical public. They would also ensure a responsible attitude to such research is taken by both industry and academia.

PROTECTION OF INTELLECTUAL PROPERTY

In recognition that the Community's industries, agricultural producers and academic researchers are in a position to be competitive at the international level, the Commission has proposed a Directive on the legal protection of biotechnological inventions. The fact that differences in the legal protection of of such inventions exist even among the Member States, and that such differences could create barriers to trade and to the creation and proper functioning of the internal market, has called for a harmonised legislation concerning the legal protection of biotechnological inventions. The harmonisation of patent protection in the Commission's proposal for a directive presents an essential element in the Community's multifaceted strategies for biotechnology. [See editors' update, p. 161.]

MONITORING DEVELOPMENTS

In conclusion, I wish to state that the European Commission continues to monitor, on a world-wide basis, developments in animal transgenesis. For example, we follow with great interest what is happening through many channels including: bilateral fora such as the US/EC Task Force for Biotechnology Research through which we monitor the USA; the 'Provisional Advisory Committee on Ethical Aspects of Genetic Modifications of Animals' which has been set up in the Netherlands through which we monitor developments in Member States; and through the OECD Group of National Experts on Safety in Biotechnology. We produce and continually update a bibliographic review of developments in this field. We listen attentively to consumer concerns and public responses, as expressed through interest groups, or measured by Eurobarometer surveys, and to the voice of industry through the European Trade Association for Advanced Animal Breeding (ETAAAB) and other associations. We have set up a mechanism – the Group of Ethical Advisors – to assess and advise the Commission. We stand ready to adapt our existing legislation or to complement it by means of standards. We have the resources to research both the socio-economic and ethical issues raised and the technical infrastructure. As with other aspects of biotechnology, we shall continue to pursue the safe and beneficial development of innovations for the health, well-being and prosperity of Europe and its peoples.

20 Political Implications of Animal Biotechnology

Hiltrud Breyer MEP

Benefits will, undoubtedly, be provided by genetic engineering, but these benefits (in the form of profits) will largely accrue to the biotechnology industry. On the other hand there are costs and risks generated as a consequence of using genetic engineering techniques. For example, farm animals and laboratory animals suffer as a consequence of genetic manipulation. Furthermore, the deliberate release of genetically manipulated life-forms entails high environmental risk due to increased habitat competition and loss of varieties. In this, the final chapter in this volume, I wish to focus on the European Union's (EU) biotechnology policy and its implications for the animal welfare movement.

BIOTECHNOLOGY POLICY

By the time the Members of the European Parliament (MEPs) debated the European Commission's Proposal for a Council Directive on the Patenting of Biotechnological Inventions (COM(89)496 – SYN 159 final) in October 1992, the European Patent Office (EPO) had already granted a patent for the genetically engineered 'oncomouse' (lucidly described, for example, by Wheale and McNally 1990a p. 6). This was the first patent to be granted for a mammal in Europe and there are many others pending applications for patents on genetically engineered animals. In practice, the bio-industry paves the way forward; European Parliamentary decisions come only after the event, merely affirming the situation which already exists. [See editors' update, p. 161.]

On the one hand, the bio-industry lobbies for the deregulation of biotechnology, whilst on the other it pressures for the regulation of ethical concerns. Thus the Senior Advisory Group on Biotechnology (SAGB), an industry pressure group composed of the major multinational biotechnology corporations including ICI and Monsanto, lobbied the European Commission to appoint a high-level external body to advise the Commission on ethical issues. The aim of this ethics committee – the Commission's Group of Advisers on the Ethical Implications of Biotechnology – is not to address the the risks of genetic engineering but to eradicate the 'bad image' that certain aspects of genetic engineering has both for policy makers and the public.

This Advisory Group of originally 6, so-called, ethical 'experts' was chosen by the Commission and meets behind closed doors. Despite the fact that it is financed by public money, the public apparently has no right to know what it is doing. Furthermore, the Commission is not focusing on reducing the risks of genetic engineering but on actions that are likely to reduce the general *fear* of risk and to create public acceptance for genetic engineering activities, products and services. In short, the public is not being given the opportunity to say no to this new technology.

The Commission states in an internal paper, '[t]he question of ethics related to biotechnology and animals is, after questions related to the human genome, the second most sensitive topic in the eyes of the general public' (SEC(91)629 final p. 11). The European Commission and the bio-industry want to divert public pressure over ethical issues like transgenic animals away from the mainstream political process by preventing public interest groups from direct political lobbying. They hope that the Commission's Group of Advisers on the Ethical Implications of Biotechnology will be a form of 'tranquiliser' for the general public. No wonder that this Group of Advisers is discussing the institutionalisation of a European Agency for Ethical Authorisation! The Commission's aim is clear: they want to control the standards of what is to be considered ethically sound. I think ethical questions should be decided by society and should not be delegated to 'closed shops' whereby ethical discussion is taken away from the general public.

Let us take, for example, the question of the patenting of life. It is not only an ethical question, it is also a political question. Are we going to let the bio-industry and the market-place decide and determine the future of the animal kingdom? Because by permitting the patenting of organisms we are providing for the privatisation of nature.

There is no reason why genetically engineered animals should be patentable. They are not an invention and therefore fail to fulfil a basic requirement of patentability. The patentability of genetically engineered animals will engender a dramatic increase in animal experimentation. Some of the resultant animals will be engineered to be abnormal at birth, and generations of animals will suffer. And if animal welfare groups demand amendments to the draft EC Directive on the patenting of life-forms such that patents will not be allowable on animals engineered to suffer, then the extended debates on this issue will only serve to divert their attention away from more general welfare concerns. In addition to which, genetic manipulation will always result in some animals suffering because its effects are difficult to control and often unpredictable. On this issue, I believe that reassurances from the bio-industry are merely false hopes and promises.

In the summer of 1992 the General Accounting Office (GAO) of the USA said that the 'oncomouse' had had no positive effect on cancer research. Although the Patent and Trademark Office in the USA (US PTO) granted the patent for the 'oncomouse' in 1988 (see Wheale and McNally 1990a pp. 6–7) by 1992 not a single cancer treatment has been developed using it. Furthermore the GAO found the patented 'oncomouse' to be counterproductive because too large a percentage of cancer research funds were being diverted to this particular approach to cancer research to the detriment of alternative approaches.

RISKS POSED BY TRANSGENICS

Another problem is that of the risks associated with transgenic animals. We know now that there is the possibility of horizontal gene transfer, that is, the transfer of genes between organisms, even between species. We have enough scientific evidence to be sure about that. Yet despite the fact that this risk has been demonstrated the European Commission still wishes to deregulate the bio-industry. In many Member States the Directives on Deliberate Release (90/220/EEC) and Contained Use of Genetically Modified Micro-Organisms (90/219/EEC), adopted in 1990, were yet to be implemented by national laws in 1992; they were not fully implemented in Germany, for example.

Directives 90/220/EEC and 90/219/EEC are themselves flawed because certain key points are missing from the legislation. For example, industry has not been required to take the full liability when an accident happens, that is, there is no 'strict liability' requirement. It should make us suspicious when industry argues that there will be no accident but at the same time strongly refuses to underwrite the liability for any accidental damage. What is industry afraid of? Is it simply that the European Commission is giving way under great pressure from the bio-industry which wants to deregulate these two Directives?

DEREGULATION AND ANIMAL WELFARE

The idea behind the drive for deregulation is to move away from 'case-by-case' assessment, and to use instead standards set by industry itself in place of formal legislation (see Wheale and McNally 1990b; 1993). Moreover, 'horizontal' and process-based legislation and regulations are being systematically eroded by the adoption of new product-based ('vertical') Directives such as the Pesticides Directive (91/414/EEC).

Under the horizontal Deliberate Release Directive (90/220/EEC) the genetically modified organisms are to be evaluated as organisms with potential to cause harm and not as products with potential utility. Therefore, under this Directive a Competent Authority can only assess those aspects of a genetically engineered organism which have a bearing on environmental risk – its capacity for survival, reproduction and dispersal. The Competent Authority is not authorised to consider its efficiency as a product, therefore Directive 90/220/EEC does not permit considerations of utility to override considerations of environmental protection; potential benefit cannot justify placing the environment at risk. This contrasts with the situation under sectoral (vertical) legislation, for example, with regard to the testing of proprietary medicinal products. Deregulation in the EC will result in a loss of environmental protection (see Wheale and McNally 1993).

In 1987 the European Commission published its Proposal for a Council Regulation on the Placing on the Market of Genetically Modified Farm Animals and Semen, Ova, and Embryos. This Regulation will not be discussed in the national parliaments of the EU Member States, which is a symptom of the huge democratic deficit in the EC. At a European level, legislative power resides not in the European Parliament but with the Council of Ministers. Thus, if the Council adopts a Regulation, it immediately becomes law in all Member States of the EU.

The draft Regulation states that genetically modified farm animals (fish, of course, are also included) and semen, ova and embryos thereof *may* have a considerable impact on agricultural productivity. We know the opposite is true, but even the European Commission only says that it *may* have an impact. If the Commission itself is not sure whether or not genetically modified farm animals will have an impact on agricultural productivity, and also states in its draft Regulation that genetically modified farm animals can have adverse effects in the environment, why is there a need of this Regulation?

The answer is that the bio-industry wants to put its products on the market, for example, genetically modified hormones for sheep that will cut the cost of shearing but which may cause abortion in pregnant sheep. The effect of such products is that they increase the use of pharmaceuticals in farming and generate tremendous risks. The long-term risks cannot even be assessed. Just as the destruction of the ozone layer became widely acknowledged only after three decades, the long-term effects of transgenic animals are not known at present.

The multinational company Merck has applied for a patent in the UK for the so-called 'macro-chicken'. It is likely that this Merck 'macro-chicken' will have a variety of health problems. The amount of suffering borne by such transgenic animals is not known because such information resides with the companies (and, of course, they would be very stupid if they were to admit to the general public: 'Yes, these animals are actually suffering!').

The draft Regulation of the European Commission does not even take the suffering of animals into account. Animal welfare groups are not involved in the regulatory process. In Germany we learned that during the authorisation process for genetically modified plants under Directive 90/220/EEC, the so-called Competent Authorities did not know what questions they should ask of industry. Therefore I am afraid that the Standing Committee to be formed as a result of this Regulation will be similarly reliant on whatever information the bio-industrial applicant happens to supply to it. The same thing will happen to the draft proposal for a European Parliament and Council Regulation on novel foods and novel food ingredients (COM(93)631). How can self-regulation by the industry be trusted to evaluate the risk of the novel foods? Whilst animal rights and welfare concerns of animals such as giant pigs are obvious to animal welfare groups, the inherent risks of genetically modified organisms are obviously not recognised by the European Commission. Therefore animal welfare groups world-wide must get involved in the debate.

SUSTAINABILITY AND BIODIVERSITY

Industry has highjacked the concepts of 'sustainability' and 'biodiversity'. To claim that genetic engineering increases biodiversity is absurd; in fact, it will lead to a further loss of biodiversity, as more and more wildernesses and wildlife are obliterated. The same petrochemical–pharmaceutical companies that are responsible for the loss of biodiversity and the destruction of sustainable agriculture and husbandry now hope to profit by making their next venture – the genetically engineered exploitation of nature – politically and publicly acceptable.

Natural biodiversity will be drastically reduced if there is political and public acceptance of genetic engineering and the patenting of animals and other living creatures. The communications of the bio-industry clearly reveal an intensifying race for corporate control over the world's potentially useful germplasm. This will inevitably lead to the loss of plants and animals. Genetic engineering decisions will determine how we live, what varieties of plants and animals we raise, what we eat, and even determine the value of life itself.

In sum, the consequences of genetic engineering for animals are tremendous, and we should not expect the institutions of the EC to react in the way animal welfare and animal rights groups would wish. My experience of the European Commission is that it only supports industry. However, the animal welfare groups do have a big influence. For example, the BST moratorium [now extended to the year 2000] shows that we can be effective if we apply adequate political pressure, especially when broad coalitions of interests are possible.

The lesson to be learned from ozone-destroying chlorofluorocarbons, which scientists and industry thought to be safe, is that, in the case of the genetic engineering of animals, this is the last opportunity we may have to get it right first time!

Discussion V

Prof. Robin Attfield, Cardiff University: I have a question for Dr Peter Wheale. Peter was resuscitating Rawls's contractualist theory to cope with transgenic animals and other animal issues. He said that if the people who were in the 'original position' were to imagine themselves to be in the position of certain animals, then those animals will be included in the principles of justice that emerged. I suggest that the trouble with this idea is that there are a lot of animals with little or no subjectivity who nobody can easily imagine themselves becoming or being, and they will get left out. So is not his way of arguing going to be inadequate for a lot of animals?

Secondly, Peter suggested that if the people who were making choices about principles had sufficient sympathies in the first place they would come up with sufficient principles in the end. I want to comment that that is fine but under such circumstances nothing would depend on the contract which is the whole core of Rawls's theory. Rawls's theory is about what people would rarely bargain for if they were self-interested in the first place. By the time you have changed that, you do not need his contracts at all and should accordingly jettison the idea that morality is a kind of social contract altogether.

Dr Peter Wheale, Bio-Information (International) Ltd, London: With regard to the first point, the issue really was over sentients. Rawls's notion of agency would not necessarily apply to any living creature which had any sort of sentience. I suggested the animal would have to be capable of some form of benefit or suffering. I did say, *some* forms of life, for example, certain mammals, when I was talking about the Rawlsian contractualistic theory: I did qualify my statement with, 'some forms of life' and I thought I made it explicit that I was referring only to sentient life. I used the analogy of non-competent humans in order to show how Rawlsian contractualism could be extended to the non-rational (non-competent) agent. The individuals in the 'original position' can allocate rights to other animals precisely because we allocate rights to human incompetents, for example severely mentally challenged people or comatosed individuals. Thus we, as advocates, accord rights to individuals and groups who cannot assert or claim them for themselves. It is in this sense that I am suggesting that rational individuals in the Rawlsian 'original position' (you can imagine these people, if you like, as people with paper bags over their heads, who are ignorant of who they will be and what social status and function they will hold subsequent to the social contract) being asked to decide what sort of justice they would have in their society, not knowing who they are or what they will become.

In relation to 'moral equilibrium', I pointed out, however, that Rawls concedes that rights will exist for people who are *not* rational agents. I therefore argued that if one (as a rational agent) can imagine what it would be like to be a severely mentally challenged person or abnormal foetus (which is the implication of Rawls's contractualist argument) then it should be possible to extend our imagination to envisage what rights within society should be allocated to monkeys or cows for example, and to certain other mammals. The basis for my assertion is that we do have information about these creatures, we do have an understanding of their habits and habitats, of what makes them suffer and what environments benefit them. And so I have argued that it is not beyond our human imagination to envisage what rights they should have if we found ourselves transported from the Rawlsian 'original position' to their position. In a just society I suggest that there should be room for advocacy in Rawls's theory of contractualism, advocacy for sentient creatures which are unable to claim their own rights for themselves.

Meredith Lloyd-Evans, BioBridge: I would like to put a question to Professor Attfield. Just how far does the essence of animals reside in single genes?

Prof. Robin Attfield: I am not a geneticist, but it is my understanding that a difference of one gene does not make a sheep cease to be a sheep. The phenotype would usually still be that of a sheep, and the animal's good would still be largely determined by its evolutionary origins and ancestry. So if the question is whether a difference of one gene makes a sheep cease to be a sheep or even to cease to be an animal, I do not think it does. But I should add that nothing follows from this about whether the transgenic manipulation of sheep is ethically acceptable. For the outcome might be a sheep lacking in some of the capacities which normal sheep have, and experiencing an impoverished quality of life. While other considerations might justify such an outcome, producing this outcome would only be justifiable if there are other considerations which actually outweigh any such loss to quality of life.

Meredith Lloyd-Evans: I was not asking the question from a genetics or technological point of view. It is not a question of genes. I was actually asking from the ethical point of view. For example, to go back to the issue of animal tissue, if you were a Muslim what would you do with a humanised piglet? What would the ethical position be then?

Dr Donald Bruce, Society, Religion and Technology Project, Church of Scotland: I wish to take issue with the view expressed by Richard Ryder in the Introduction to these contributions. If I understand it correctly, Richard Ryder's view is that we should concern ourselves only with whether or not the animal suffers rather than its 'intrinsic value'.

Richard Ryder, RSPCA: I find it difficult to understand what is meant by 'value'. What constitutes the value *is* the animal's consciousness.

Prof. Robin Attfield: I think we do not need to get into great difficulties about the word 'value'. It strikes me that we have all been talking about the word 'value' all along. When people say 'intrinsic value' they just mean this to be a starting point. So, in fact, presumably Richard Ryder is saying that pain and suffering are of negative value. With 'intrinsic value' we want to say that there is more to it than just happiness.

Jose MacDonald: This is a question for Dr Lex. The industry most affected by biotechnology is going to be agriculture. I do not see any farmer or farming representative on any of the ethical committees. We have a headmistress of a girls' school (Dame Mary Warnock) advising on ethics and biotechnology [on the Commission's Group of Advisers] and no stock farmer with his feet in the mud dealing with the actual animals. What is your view on this situation?

Dr Maurice Lex, DG XII of the European Commission: DGVI, the European Commission's Directorate-General for Agriculture, is fully represented on the biotechnology committees. It usually sends two people, who are very, very powerful individuals, to speak up for agriculture, but I really do not know whether they consult farmers. I am convinced they must do so on some of their committees.

I must take issue with Hiltrud Breyer's accusation that the European Commission always does what industry says, because we listen to all sides and, in fact, just last week there was a Round Table where environmental groups, as well as industry, were represented, and David Williamson, the Secretary-General of the Commission's Biotechnology Coordination Committee (BCC), confirmed that he wants to hear the views and listen to people generally, as to what they want.

Hiltrud Breyer, MEP: I was there at this meeting to which Mr Lex refers. However, despite being the Parliament's official Rapporteur on the European Commission's Communication on Biotechnology policy (SEC(91)629), I was not invited because you had chosen some other member of European Parliament which you thought better fits your interests.

To illustrate my point about the Commission I should like to mention a booklet whose publication and distribution was financed with public funds. When it first came out the booklet was sponsored by Eli Lilly, the big USA company. The European Commission took up this booklet and even amended it because Eli Lilly had at least included some critical points about genetic engineering, for example that genetic engineering 'can harm', or 'does harm'; such comments were removed by the Commission. The printing and distribution of this booklet was then funded by the Commission.

I do not know of any non-governmental organisations (NGOs), for example, animal welfare rights groups, who are subsidised by the Commission to publish and distribute their literature!

Ruth McNally, Bio-information (International) Ltd, London: I have studied the documents on EC biotechnology policy sent to the Commission by industry's Senior Advisory Group for Biotechnology (SAGB), a lobby group for the largest biotechnology companies operating in Europe. I have also undertaken a detailed study (see Wheale and McNally 1993) of the Commission's subsequent Communication on biotechnology policy (SEC(91)629). There is an extraordinary similarity between the SAGB's criticisms of, and suggestions for, EC biotechnology policy and the subsequent European Commission's statement on competition policy for biotechnology in the EC. Indeed, the resemblance was even commented on in the Commission's own European Biotechnology Information Service (EBIS) Newsletter. Such similarity does suggest that the Commission is significantly influenced by bio-industry (or at least the major bio-industry companies).

On the subject of the Commission's Group of Advisers on Bioethics, I remember reading in the EBIS Newsletter that the Biotechnology Coordination Committee (BCC) considered that it should comprise a small group of people of 'high moral standing' (EBIS July 1991 p. 3). I would like to ask Dr Lex how the Commission selected 6 such persons, and exactly what status does the Group have in respect of Commission activity? I would also like to ask: 'Why 6?' After all, there are *12* Member States. [In 1994 the group was expanded to 9.]

Dr Maurice Lex: I will do my best to answer this question but I was not responsible for the appointment of that Committee. Just like you, I also asked: 'Why 6?' And the answer given was that if it was too large a forum it would be very difficult to get any consensus viewpoint, but the decision that there should be only 6 people was taken at a very high level.

Ms Breyer stated that the Commission's Advisory Group on Bioethics operates behind closed doors. Well, that is true, but I think that that was what the Group itself asked for. The Group members wanted to be able to express their views freely and therefore they wanted to operate in private. But the Commission has said that it will make public any report that is actually produced by that Group and the Group does have a tremendous amount of freedom to raise any issues that it wishes, and present them to the Commission.

As to the booklet which we produce, *Biotechnology for All*, I am told in the Commission that we always have to be very nice to MEPs, but I must contradict you, Ms Breyer, when you claim that it was an Eli Lilly publication. It was actually first produced by the Department of Trade and Industry (DTI) here in the UK. They were the first people to produce it, although Eli Lilly money was used in the production with the DTI. We saw the publication and we thought that it was a very good one, particularly at the level of school children and therefore we asked the DTI if we could use it. Why is it that although people write about this publication in magazines and other places, they never write to us to contradict what is in *Biotechnology For All*? It appears to be a very popular publication and certainly no-one has ever written in to say that it is factually incorrect. If it is, we would like to hear this sort of comment please. Thank you.

Geraldine O'Brien, Cork Environmental Alliance, Ireland: I would like to ask Dr Lex if he could justify, defend or explain whether one of the terms of reference of the Commission's Advisory Group on Bioethics is to improve public understanding and acceptance of biotechnology, because to me that just sounds like a 'rubber stamp' operation. Also I want to know whether the deliberations of this Group are going to be publicly available under the Freedom of Access to Information on the Environment, Directive (COM(88)484) which comes into force in January 1993?

Dr Maurice Lex: It was said that the Commission's Advisory Group on Bioethics is just a 'tranquilliser' for the general public. I honestly do not take that sceptical view of the Commission. I believe that the Bioethics Group was set up in all seriousness to look at these ethical issues, in the same way as every Member State has set up ethical committees. The Nuffield Council in the UK has set up an ethical committee; and the Ministry of Agriculture Fisheries and Food (MAFF) has set up an Ethical Group. They are talking with one another. There is openness, there is frankness, but, as with, I think, other ethical committees in Europe, they have said that they want to debate things in private, and I think that they are perfectly entitled to do that because that enables them to speak honestly, and we will then not associate any particular thought or approach with any particular individual within that Committee or Group. As an independent body such as it is, the Advisory Group is not dictated to by the Commission; I can assure you, its Chairman is an MEP.

Hiltrud Breyer: That is not true! I asked him 2 days ago: and he said: 'I am not there as an MEP. I am there as a private person, and I would be staying there even if I was not an MEP', so do not misinform us please.

Dr Maurice Lex: I am sorry if I misled you. Those individuals are there as individuals. They are not representing any organisation. They are there as individuals, I must stress that.

Geraldine O'Brien: I do not think you answered my question. The question was, is the function of the Advisory Group on Bioethics to defend and improve public understanding and acceptance of genetic engineering?

Dr Maurice Lex: Yes. Of course, I did not draft those terms of reference of the Bioethics Group, but there is a concern that there is confused thinking amongst the public. A lot of people lay tremendous emphasis on the risks of biotechnology. I must say that the risks are totally unproven: all the evidence points the other way. Now, Ms Breyer also said that we are not spending research money on looking at those risks. We have committed 10 million ECU (MECU) to studying the bio-safety aspects of biotechnology. There is a report of a meeting held in Goslar, in Germany, in 1992 when scientists from all over the world assembled to look at the risks of biotechnology. These things are being looked at very seriously.

Hiltrud Breyer: I think we are too cautious about speaking out about animal suffering. The scientists say they will not suffer. Do you think that is not exaggerating also? Not a risk? And you say there will be a product in 2 years! How can you be sure? You say that the evidence for potential benefits is that industry spends a lot of money on it. But if, for example, we spend money on military arms, it does not mean that there will be a war. Nor does it mean that military arms are good because industry spends money developing them.

I would like to make just a brief comment about the risk. Of course the European Commission says it conducts risk research, but what do they make of it? Suppose it shows there is, for example, evidence of horizontal gene transfer. The European Union never will use this evidence to alter regulating legislation. The risk research sponsored by the

European Commission is mainly used to undertake market research; for example, a lot of the 3 per cent of the research budget for the Biotechnology Programme which you cited as being to assess ethical and socioeconomic effects and technological risks is misused for doing market research.

And just one final example of why I do not trust industry. The existence of horizontal gene transfer has been denied by industry, but when I asked industrial companies to put an identifying genetic 'marker' in their transgenic organisms so that they can be held accountable and liable for adverse consequences, do you know what they say to me? 'We cannot because there is risk of horizontal gene transfer of the genetic marker itself.' So, how can this contradiction make sense?

References

ACNFP (1992) *Advisory Committee on Novel Foods and Processes, Report of Fourteenth Meeting,* 13 February (London: Department of Health).

ACOST (1990) *Developments in Biotechnology, Advisory Council on Science and Technology* (London: HMSO).

Ali, S. and Clark, A.J. (1988) Characterisation of the gene encoding ovine β-lactoglobulin, *J. Mol. Biol.*, 199, 415–26.

Ali, S. et al. (1990) Characterisation of the alleles encoding ovine β-lactoglobulins A and B, *Gene*, 91, 201–7.

Allen, N.D. *et al.* (1988) Transgenes as probes for active chromosomal domains in mouse development, *Nature*, 333, 852–5.

Al-Shawi, R. *et al.* (1990) Expression of a foreign gene in a line of transgenic mice is modulated by a chromosomal position effect, *Mol. Cell. Biol.*, 10, 1192–8.

Anderson, R. (1986) Rabies control: Vaccination of wildlife reservoirs, *Nature*, 322, 304–5.

Anon (1992) Tales of the unexpected, *Lancet*, 339, 298.

Archibald, A.L. *et al.* (1990) High level expression of biologically active human α1-antitrypsin in transgenic mice, *Proc. Natl Acad. Sci.* USA, 87, 5178–82.

Avery, N. *et al.* (1993) *Cracking the Codex: An Analysis of Who Sets World Food Standards* (London: National Food Alliance).

Bacon, F. (1624) *The New Atlantis*, A.B. Gough (ed.) (Oxford: Oxford University Press, 1924).

Barinaga, M. (1992) Knockout mice offer first animal model for CF, *Science*, 257, 1046–7.

Baxby, D. (1985) Vaccinia virus, in G.V. Quinnan (ed.), *Vaccinia Viruses as Vectors for Vaccine Antigens* (New York: Elsevier), 3–7.

Baxby, D. *et al.* (1986) Ecology of orthopoxviruses and use of recombinant vaccinia vaccines, *Lancet*, 11 October, 850–1.

BCC (1992) *Biotechnology after the 1991 Communication: A Stock Taking* (Luxembourg: Office for Official Publications of the European Communities).

Behringer, R.R., *et al.* (1988) Heart and bone tumors in transgenic mice, *Proc. Natl Acad. Sci.* USA, 85, 2646–52.

Behringer, R.R. *et al.* (1989) Synthesis of functional hemoglobin in transgenic mice, *Science*, 245, 971–3.

Bizley, R.E. (1994) Patenting Animals, *The Genetic Engineer and Biotechnologist*, 14, 41–7.

Blindell, T. (1992) *Daily Telegraph* Editorial, 13 March.

Bolt, D. *et al.* (1988) Improved animal production through genetic engineering: transgenic animals, *Proceedings of forum Veterinary Perspectives on Genetically Engineered Animals*, AVMA, 58–61.

Bondioli, K. (1992) Commercial cloning of cattle by nuclear transfer, in G.E. Seidel Jr (ed.) *Symposium on Cloning Mammals by Nuclear Transplantation*, Colorado State University, 33–8.

Bonnerot, C. *et al.* (1990) Patterns of expression of position dependent integrated transgenes in mouse embryos, *Proc. Natl Acad. Sci.* USA, 87, 6331–5.

Boyle, C. *et al.* (1986) *People, Science and Technology: A Guide to Advanced Industrial Society* (London: Harvester-Wheatsheaf).

Brinster, R.L. *et al.* (1984) Transgenic mice harboring SV40 T-antigen genes develop characteristic brain tumours, *Cell*, 37, 367–79.

Brinster, R.L. *et al.* (1988) Introns increase transcriptional efficiency in transgenic mice, *Proc. Natl Acad. Sci.* USA, 85, 436–40.

Brochier, B. *et al.* (1991) Large-scale eradication of rabies using recombinant vaccinia-rabies vaccine, *Nature*, 354, 520–2.

Brundtland, G. (1987) *Our Common Future* (Paris: United Nations World Commission on Environment and Development).

Bulfield, G. (1990) Genetic manipulation of farm and laboratory animals, in P. Wheale and R. McNally (eds) *The Bio-Revolution: Cornucopia or Pandora's Box?* (London: Pluto Press), 18–23.

Bull, A.T. *et al.* (1982) *Biotechnology: International Trends and Perspectives* (Paris: OECD).

Callicott, B.J. (1992) *Animal Liberation: A Triangular Affair* republished in *The Animal Rights/Environmental Ethics Debate*, E. C. Hargrove (ed.) (University of New York Press).

Caplan, A.L. (1985) Ethical issues raised by research involving xenografts, *JAMA*, 254, 3339–43.

Cartwright, W. (1991) The ethics of xenografting in Man, in W. Land and J.B. Dossetor (eds) *Organ Replacement Therapy: Ethics, Justice and Commerce* (Berlin: Springer-Verlag).

Clark, J. (1991) *Farmer's Weekly* Editorial, 29 March.

Clark, L.L. *et al.* (1992) Defective epithelial chloride transport in a gene-targeted mouse model of cystic fibrosis, *Science*, 257, 1125–8.

Collins, F.S. and Wilson, J.M. (1992) A welcome animal model, *Nature*, 358, 708–9.

COM(88)484 Final *Proposal for a Council Directive on the Freedom of Access to Information on the Environment.*

COM(88)496 Final – SYN 159 *Proposal for a Council Directive on the Legal Protection of Biotechnological Inventions.*

COM(93)631 Final *Proposal for a European Parliament and Council Regulation on Novel Foods and Novel Food Ingredients.*

Dumbell, K.R. (1985) Aspects of the biology of orthopoxviruses relevant to the use of recombinant vaccinia as field vaccines, in G.V. Quinnan (ed.), *Vaccinia Viruses as Vectors for Vaccine Antigens* (New York: Elsevier), 9–13.

Durning, A. (1992) *How Much Is Enough?* (London: Earthscan).

Dworkin R. (1970) The law relating to organ transplantation in England, *Medical Law Review*, 353, 355–59.

Dworkin R. (1977) *Taking Rights Seriously* (Cambridge, MA: Harvard University Press).

86/609/EEC *Council Directive on the Approximation of Laws, Regulations and Administrative Provisions of the Member States Regarding the Protection of Animals Used for Experimental and Other Scientific Purposes.*

87/22/EEC *Council Directive on the Approximation of National Measures Relating to the Placing on the Market of High-Technology Medicinal Products, Particularly those Derived from Biotechnology.*

89/342/EEC *Council Directive Extending the Scope of Directives 65/65/EEC and 75/319/EEC and Laying Down Additional Provisions for Immunological Medicinal Products Consisting of Vaccines, Toxins or Serums and Allergens.*

89/381/EEC *Council Directive Extending the Scope of Directives 65/65/EEC and 75/319/EEC on the Approximation of Provisions Laid Down by Law, Regulations or Administrative Action Relating to Proprietary Medicinal Products and Laying Down Special Provisions for Medicinal Products Derived from Human Blood or Human Plasma.*

89/455/EEC Council Decision of 24 July 1989 introducing Community measures to set up pilot projects for the control of rabies with a view to its eradication or prevention, *Official Journal of the European Communities*, No. L 233, 2 August, 19–21.

90/219/EEC *Council Directive on the Contained Use of Genetically Modified Micro-Organisms.*

90/220/EEC *Council Directive on the Deliberate Release of Genetically Modified Organisms.*

90/425/EEC *Council Directive Concerning Veterinary and Zootechnical Checks Applicable in Intra-Community Trade in Certain Live Animals and Products with a View to the Completion of the Internal Market.*

90/679/EEC *Council Directive on the Protection of Workers from the Risks Related to Exposure to Biological Agents at Work.*

91/414/EEC *Council Directive Concerning the Placing of Plant Protection Products on the Market.*

Ekesbo I. (1992) Biotechnology for control of growth and product quality in meat production: implications and acceptability for animal safety, health and welfare, unpublished paper presented to the international symposium on *Biotechnology for Control of Growth and Product Quality in Meat Production: Implications and Acceptability* at Wageningen Agricultural University.

Esposito, J.J. and Murphy, F.A. (1989) Infectious recombinant vectored virus vaccines, *Advances in Veterinary Science and Comparative Medicine*, 33, 195–247.

European Commission (1993) *Growth, Competitiveness, Employment: The Challenges and Ways Forward into the 21st Century* (White Paper) (Luxembourg: Office for Official Publications of the European Community).

FAC (1990) *Food Advisory Committee Report on its Review of Food Labelling and Advertising 1990*, FAC/Rep10, MAFF (London: HMSO).

FAWC (1988) *Report on Priorities in Animal Welfare Research and Development* (Tolworth: FAWC).

Field, L. (1988) Atrial natriuretic factor–SV40 T antigen transgenes produce tumors and cardiac arrythmias in mice, *Science*, 239, 1029–32.

First, N.L. (1990) New animal breeding techniques and their application, *Proceedings of the Second Symposium on Genetic Engineering of Animals*, published in *Journal of Reproduction and Fertility*, supplement no. 41, 3–14.

Flamand, A. *et al.* (1990) Monitoring the potential risk linked to the use of modified live viruses for antirabies vaccination of foxes, in I. Economidis (ed.), *Biotechnology R&D in the EC – BAP: Part I* (Brussels: Commission of the European Communities).

Flamand, A. *et al.* (1992) Eradication of rabies in Europe, *Nature*, 360, 115–16.

Food Magazine (1992), 2:19, 1, 10–11.

Food Magazine (1993) Pure food campaign, Washington DC, USA, 1992, 2:20, 3.

Fox, M. (1990) Transgenic animals: Ethical and animal welfare concerns, in P. Wheale and R. McNally (eds) *The Bio-Revolution: Cornucopia or Pandora's Box?* (London: Pluto Press), 166–74.

Francione, G.L. (1990) Xenografts and animal rights, *Transplantation Proceedings*, 22, 1044–6.

Frey, R.G. (1983) *Rights, Killing and Suffering* (Oxford: Blackwell), 27–9.

Fromm, E. (1970) *The Crisis of Psychoanalysis* (Harmondsworth: Penguin).

GATT Final Act (1991) (the Dunkel draft), MTN. TNC/W/FA, (Geneva: GATT), 20 December.

Gene Exchange (1994) Rhone Merieux seeks a license for recombinant rabies vaccine, *Gene Exchange*, 4, 8–9.

Griffin, R. (1992) *Animal Minds* (Chicago: University of Chicago Press).

Grosveld, F. *et al*, (1987) Position independent, high level expression of the human beta-globin gene, *Cell*, 51, 975–85.

Haerlin, B. (1990) Genetic engineering in Europe, in P. Wheale and R. McNally (eds), *The Bio-Revolution: Cornucopia or Pandora's Box?* (London: Pluto Press), 253–61.

Harris, S. *et al.* (1991) Developmental regulation of the sheep β-lactoglobulin gene in transgenic mice, *Devel. Genet.*, 12, 299–307.

Hobsbawm, E. and Rudé, G. (1973) *Captain Swing* (Harmondsworth: Penguin).

Hodgson, J. (1992) Whole animals for wholesale protein production, *Biotechnology*, 10, 863–6.

Holland, A. (1990) The biotic community: A philosophical critique of genetic engineering, in P. Wheale and R. McNally (eds) *The Bio-Revolution: Cornucopia or Pandora's Box?* (London: Pluto Press), 166–74.

Home Office (1992) *Statistics of Scientific Procedures on Living Animals*, (London: HMSO).

Hooper, M. *et al.* (1987) HPRT-deficient (Lesch–Nyhan) mouse embryos derived from germline colonisation by cultured cells, *Nature*, 326, 292–5.

Jackson, C. (1992) Mad dogs and Englishmen, *Kangaroo News*, October, 13.

Jahoda, M. (1982) Once a jackass ..., *Nature*, 295, 173–4.

Jenkins, R. (1991) *Elements for an Evaluation of Food Biotechnology from a Consumer's Point of View* (Brussels: European Commission, Consumer Policy Services) November, 237.

Jenkins, R. (1992) *Bringing Rio Home* (London: SAFE Alliance).

Jonas, H. (1984) *The Phenomenon of Life* (Chicago: University of Chicago Press).

Jorgensen, R. (1990) Altered gene expression in plants due to trans interactions between homologous genes, *Trends in Biotechnology*, 8, 340–4.

Kaplan, C. (1989) Vaccinia virus: A suitable vehicle for recombinant vaccines?, *Archives of Virology*, 106, 127–39.

Klinger, T. *et al.* (1991) Radish as a model system for the transfer of engineered gene escape rates via crop–weed mating, *Conservation Biology*, 5, 531–5.

Kramer, D. and Templeton, J. (1988) Genetically engineered disease resistance in mammals, *Proceedings of forum Veterinary Perspectives on Genetically Engineered Animals*, AMVA, 48–53.

Kuehn, M.R. *et al.* (1987) A potential animal model for Lesch–Nyhan syndrome through introduction of HPRT mutations into mice, *Nature*, 326, 295–8.

Kukathas, C. and Pettit, P. (1990) *Rawls: A Theory of Justice and its Critics* (Cambridge: Polity Press).

Lang, T. (1992) *Food Fit for the World?, Sustainable Agriculture, Food and Environment* (London: SAFE Alliance).

Linzey, A. (1990) Human and animal slavery: A theoretical critique of genetic engineering, in P. Wheale and R. McNally (eds) *The Bio-Revolution: Cornucopia or Pandora's Box?* (London: Pluto Press), 175–89.

Lovelock, J. (1988) *The Ages of Gaia* (Oxford: Oxford University Press).

Luhmann, N. (1986) The autopoiesis of social systems, in F. Geyer and J. van der Zouen (eds), *Sociocybernetic Paradoxes: Observation, Control and Evolution of Self–Steering Systems* (London: Sage).

Macer, D. (1989) Uncertainties about 'Painless' animals, *Bioethics*, 3, 226–35.

Mackie, J.L. (1984) Can there be a right–based moral theory? in J. Waldron (ed.) *Theories of Rights* (Oxford: Oxford University Press), ch. 8.

MAFF (1991a) *Food Labelling Survey* (London: HMSO).

MAFF (1991b) *Food Safety Directorate*, News Release, *FSD*, 5/91, 17 January.

MAFF (1992) *Study Group on Ethics of Genetically Modified Foods*, News Release, no. 306, 25 September (London: MAFF).

Mahon, K.A. *et al.* (1987) Oncogenesis of the lens in transgenic mice, *Science*, 235, 1622–8.

McFall v. *Shimp* The Court of Common Pleas, Alleghery County, Pennsylvania, Order, 26 July 1978.

McNally, R. (1994) Genetic madness: The European rabies eradication programme, *The Ecologist*, 24, 207–12.

McNally, R. (1995) Doubts on EU genetic safety, *The Ecologist*, 25.

McNeish, J.D. *et al.* (1988) Legless, a novel mutation found in PHT1-1 transgenic mice, *Science*, 241, 837–9.

Morgan, J.A.W. (1990) Genetic engineering of microorganisms: free release into the environment, *58th Annual Report of the Freshwater Biological Association*, (Ambleside: FBA), 91–107.

Murphy, D. *et al.* (1987) Mice transgenic for a vasopressin–SV40 hybrid oncogene develop tumors of the endocrine, pancreas and the anterior pituitary, *Am. J. Pathology*, 129, 552–66.

Nancarrow, C.D. *et al.* (1991) Expression and physiology of performance regulating genes in transgenic sheep, *Journal of Reproduction and Fertility Symposium*, 43, 277–91.

New York Times (1990) 28 April, 9.

Owen, J. (1992) *Farmers Weekly*, Editorial, 13 March.

Palmiter, R.D. and Brinster, R.L. (1986) *Annu. Rev. Genet.*, 20, 465–99.

Palmiter, R.D. *et al.* (1991) Heterologous introns can enhance expression of transgenes in mice, *Proc. Natl Acad. Sci.* (USA), 88, 478–82.

Parfit, D. (1984) *Reasons and Persons,* (Oxford: Clarendon Press), 487.

Pastoret, P.P. *et al.* (1988) Viruses from the ecological viewpoint, Conference paper.

Pastoret, P.P. *et al.* (1992) Development and deliberate release of a vaccinia–rabies recombinant virus for the oral vaccination of foxes against rabies, in M.M. Binns and G.L. Smith (eds) *Recombinant Poxviruses* (London: CRC Press), ch. 5.

Pearce, D. (1991) *Blueprint 2: Greening the World Economy* (London: Earthscan).

Pocar, V. (1990) Animal rights: a socio-legal perspective, *Journal of Law and Society,* 19, 214–30.

Postma, O. (1989) *Daily Telegraph* Editorial, 18 August.

Pursel, V. *et al.* (1987) High level synthesis of a heterologous milk protein in the mammary glands of transgenic swine, *Proc. Natl Acad. Sci.* USA, 1696–1700.

Pursel, V.G. *et al.* (1989) Genetic engineering of livestock, *Science,* 244, 1281–8.

Rawls, J. (1972) *A Theory of Justice,* (Oxford: Oxford University Press).

Rawls, J. (1985) A Kantian Conception of Equality, in J. Rajchman and C. West (eds) *Post-Analytic Philosophy* (New York: Columbia University Press), 201–13.

Raz, J. (1986) *The Morality of Freedom* (Oxford: Oxford University Press).

Redfield, R.R. *et al.* (1987) Disseminated vaccinia in a military recruit with human immunodeficiency virus (HIV) disease, *N. Engl. J. Med.,* 316, 673–6.

Reemtsma, K.L. (1990) Ethical aspects of xenotransplantation, *Transplantation Proceedings,* 22, 1042–3.

Regan, T. (1983) *The Case for Animal Rights* (London: Routledge and Kegan Paul).

Regina v. *Dudley and Stephens* (1884), 14 QBD, 273.

Reid, W.V. (1992) Conserving life's diversity, *Environmental Science and Technology,* 6, 1090–5.

Rexroad, C.E. *et al.* (1990) Insertion, expression and physiology of growth–regulating genes in ruminants, *J. Reprod. Fert.,* supplement no. 41, 119–24.

Robertson, J. (1976) Organ donations by incompetents and the substituted judgment doctrine, *Columbia Law Review,* 48, 57–63.

Rollin, B.E. (1981) *Animal Rights and Human Morality* (New York: Prometheus Books).

Rollin B.E. (1988) An ethical perspective on genetically engineered animals, *Proceedings of forum Veterinary Perspectives on Genetically Engineered Animals,* AMVA, 35–43.

Royal Commission on Environmental Pollution (RCEP) (1989) *13th Report on the release of genetically engineered organisms to the environment* (London: HMSO).

Ryder, R.D. (1990) Pigs *will* fly, in P. Wheale and R. McNally (eds) *The Bio-Revolution: Cornucopia or Pandora's Box?* (London: Pluto Press), 166–74.

Salter, D.W. (1988) Genetically engineered disease resistance in poultry, *Proceedings of forum Veterinary Perspectives on Genetically Engineered Animals*, AMVA, 44–7.

Scanlon, T.M. (1985) Contractualism and utilitarianism, in J. Rajchman and C. West (eds) *Post-Analytic Philosophy* (New York: Columbia University Press), 215–43.

Schanbacher, F.L. (1988) Molecular farming: current status and prospects, *Proceedings of forum Veterinary Perspectives on Genetically Engineered Animals*, AMVA, 54–7.

SEC(91)629 Final *Promoting the Competitive Environment for the Industrial Activities Based on Biotechnology Within the Community.*

Shiva, V. (1992) The MTO, TRIPS and TRIMS, *Paper for the Meeting of NGOs on GATT*, Munich, 8 July.

Simons, J.P. *et al.* (1987) Alteration of the quality of milk by expression of sheep β-lactoglobulin in transgenic mice, *Nature*, 328, 530–2.

Singer, P. (1977) *Animal Liberation* (London: Paladin).

Snouwaert, J.N. *et al.* (1992) An animal model for cystic fibrosis made by gene targeting, *Science*, 257, 1083.

Stice, S.L. (1992) Multiple generation bovine embryo cloning, in G.E. Seidel Jr (ed.) *Symposium on Cloning Mammals by Nuclear Transplantation*, Colorado State University, 28–32.

STOA (1992) *Bioethics in Europe* (Brussels: European Parliament).

Storb, U. *et al.* (1986) Transgenic mice with mu and kappa genes encoding antiphosphorylcholine, *J. Exp. Med.*, 164, 627–41.

Street, P. (1992) Quoted from text of 'Fast life in the food chain', *BBC Horizon*.

Stubbings, R. (1990) *Shropshire Star* 3 July, quotation

Sumner, L. W. (1990) *The Moral Foundation of Rights* (Oxford: Clarendon Press).

Supermarketing (1992), 7 August, 28.

Tiedje, J.M. *et al.* (1989) The planned introduction of genetically engineered organisms: ecological considerations and recommendations, *Ecology*, 70, 298–315.

Tynan, J.L. *et al.* (1990) Low frequency of pollen dispersal from a field trial of transgenic potatoes, *J. Genet. Breed.*, 44, 303–6.

Umbeck P.F. *et al.* (1991) Degree of pollen dispersal by insects from a field trial of genetically engineered cotton, *J. Econ. Entomol.* 84, 1943–50.

Webster, J. (1990) Animal Welfare and Genetic Engineering, in P. Wheale and R. McNally (eds) *The Bio-Revolution: Cornucopia or Pandora's Box?* (London: Pluto Press), 24–30.

Went, D.F. and Stranzinger, G. (1990) *Experientia*, 47, 934–6.

Wheale, P. and McNally, R. (1988) *Genetic Engineering: Catastrophe or Utopia?* (London: Wheatsheaf).

Wheale, P. and McNally, R. (eds) (1990a) *The Bio-Revolution: Cornucopia or Pandora's Box?* (London: Pluto Press).

Wheale, P. and McNally, R. (1990b) Genetic engineering and environmental protection: A framework for regulatory evaluation, *Project Appraisal*, 5, 23–38.

Wheale, P. and McNally, R. (1993) Biotechnology policy in Europe: A critical evaluation, *Science and Public Policy*, 20, 261–79.

Wheale, P. and McNally, R. (1994) What 'bugs' genetic engineers about bioethics: the consequences of genetic engineering as post-modern technology, in A. Dyson and J. Harris (eds) *Ethics and Biotechnology* (London: Routledge), ch. 10, 179–201.

Whitelaw, C.B.A. *et al.* (1991) Targeting expression to the mammary gland; intronic sequences can enhance the efficiency of gene expression in transgenic mice, *Transgenic Res.*, 1, 3–13.

Wiktor, T. *et al.*, (1988) Rabies vaccine, in S.A. Plotkin and E.A. Mortimer (eds) *Vaccines* (Philadelphia: W.B. Saunders Co.).

Wilson, J. (1992) Rabies control, *Report of the 88th Session of the Intergroup on Animal Welfare*, 9 July, 4–5.

Wright, G. *et al.* (1991) High level expression of active human alpha–1–antitrypsin in the milk of transgenic sheep, *Biotechnology*, 9, 830–4.

Zimen, E. (1980) *The Red Fox* (The Hague: Junk Books).

Glossary

Note: cf., (confer) compare; Gr., Greek; L., Latin.

Allele/allelomorph: (Gr. *allelon*, one another; *morphe*, form.) Alternative forms of a gene for a given trait.

Allografting: (Gr. *allos*, other.) The transplantation of an organ or tissue between individuals of the same species; also known as *homograft* (Gr. *homos*, same).

Amino acids: The building blocks of protein structure; 20 different amino acids are commonly found in living organisms.

Amniocentesis: A prenatal screening procedure, usually carried out at around 17 weeks of pregnancy, in which a few millilitres of the amniotic fluid and floating cells surrounding the foetus are withdrawn through a needle inserted via the mother's abdomen and uterine wall.

Anthropocentric: (Gr. *anthropos*, a man; L. *centrum*, centre.) Taking mankind as the pivot of the universe.

Antibiotic: (Gr. *ante*, against or opposite; *bios*, life.) A substance capable of killing or preventing the growth of a microorganism; can be produced by another microorganism or synthetically.

Antibody: A protein produced in the body in response to the presence of a foreign chemical substance or organism (antigen), shaped to fit precisely to the antigen and in such a way as to annul its action or help to destroy it; part of the body's defence (immune) system.

Antigen: A substance which causes the immune system of the body to manufacture specific antibodies that will react with it.

Artificial insemination (AI): The application of sperm to an unfertilised egg by means other than ejaculation during sexual intercourse.

Assay: A technique that measures a biological response, for example, a plaque assay measures the number of infective viruses suspended in a measured volume of medium (see also Plaque).

Attenuated vaccine: the production of a vaccine (see also Vaccine) where the pathogenic microorganism has reduced virulence after several generations of culture *in vivo*.

Autosome: A chromosome other than a sex chromosome.

Bacteria: (Gr. *bacterium*, a little stick.) The simplest organisms that can reproduce unaided; a class of single-celled microorganism found living freely in water, the soil and in the air, and as parasites within plants and animals; *Escherichia coli* (*E. coli*) is the species commonly used as a host cell in recombinant DNA work.

Baculovirus: (L. *baculum*, rod.) Rod-shaped DNA virus which is believed to infect only the cells of invertebrate animals.

Base: Part of the building blocks of nucleic acids, the sequence of which encodes genetic information; cytosine (C), guanine (G), adenine (A) and thymine (T) are the bases in DNA; C, G, A and uracil (U) are the bases in RNA (see DNA Base Pair; Nucleotide).

Biochemistry: The study of the chemistry of living things.

Biodiversity: An ecological term which refers to all species of plants, animals and microorganisms in the world and their various ecosystems.

Biogenetic waste: Biological waste that contains genetically modified organisms; includes sewage, refuse and effluent from biotechnological processes.

Biological containment: The use of organisms in genetic engineering applications and research which are genetically engineered so as to minimise their ability to survive, persist or replicate; also applies to the use of genetically deficient cloning vectors, which are deficient in their ability to move to a new host strain; also known as 'genetic enfeeblement' and 'crippling'.

Biologicals/biologics: Term used to describe drugs which are based on substances found in living animals, for example, insulin.

Bioprocess: A process that uses complete living cells or their components, such as enzymes, to provide goods or services, for example, brewing.

Bioreactor: Vessel in which a bioprocess takes place, for example, a fermenter.

Biosphere: (Gr. *bios*, life; *sphaira*, a ball or a globe.) All life that is encompassed under the vault of the sky; the part of the earth that is inhabited by living organisms; the Earth's surface and the top layer of the hydrosphere (water layer) have the greatest density of living organisms.

Biota: The fauna and flora of a region.

Biotechnology: After Bull *et al.* (1982), the application of scientific and engineering principles to the processing of materials by biological agents to provide goods and services.

Cartesian: Pertaining to the French philosopher Rene Descartes (1596–1650), or his philosophy; Cartesian philosophy expounds mind–body dualism, ethical dualism and explanatory dualism.

Cell: (L. *cella*, a little room, from *cello*, hide.) In 1655 Robert Hooke (1625–1702), curator of the Royal Society, used the term 'cell' to describe the small, closed cavities he found upon microscopic examination of the outer bark of an oak tree; although the structures he observed were actually cell walls, which are absent from animal cells, the term has persisted to describe the basic unit of structure of all living organisms excluding viruses; as defined by Max Schultze (1825–74) a cell is 'a lump of nucleated protoplasm'; a generalised description of a cell is a mass of jelly-like cytoplasm, contained within a semipermeable membrane, and containing a spherical body called the nucleus; a single cell constitutes the entire organism of a single-celled creature such as a bacterium; a human being is composed of millions of cells.

Cell fusion: The fusing together of two or more cells to produce a single hybrid cell (see also Hybridoma technology and Monoclonal antibodies).

Chimaera: (L. *chimaera,* a mythological monster with the head of a lion, the body of a goat, and the tail of a dragon, vomiting flames.) An organism, cell or molecule (for example, DNA) constructed from material from two different individuals, or species.

Chorionic villus sampling (CVS): A prenatal screening procedure whereby cells of the chorion – the membrane which surrounds the embryo – are withdrawn for genetic analysis; the cells of the chorion are derived from the fertilised egg and contain the same genetic information as those of the foetus; CVS can be performed at any stage of pregnancy from about 8 weeks of gestation onwards.

Christmas disease: see Haemophilia.

Chromosomes: (Gr. *chroma,* colour; *soma,* body.) Darkly staining structures, composed of DNA and protein, which bear and transmit genetic information; they are found in the nucleus of eukaryotic cells and free in the cytoplasm in the cells of prokaryotes; the number of chromosomes in each somatic cell is characteristic of the species; in human beings the normal chromosomal constitution is 22 pairs of autosomal chromosomes, and one pair of sex chromosomes.

Clone: A collection of genetically identical molecules, cells or organisms which has been derived (asexually) from a single common ancestor.

Cloning: Making identical copies of biological entities – molecules, cells or individuals; hybridoma technology is a cloning technique.

Codon: Three successive nucleotides (or bases) which specify a particular amino acid or a 'punctuation mark' in the genetic code.

Colony (of microorganisms): A dense mass of microorganisms produced asexually from an individual microorganism.

Congenital disorder: A malfunction which is present at birth; the term describes all deformities and other conditions that are present at birth whether they are inherited or newly arisen as a result of adverse environmental factors or transgenic manipulation of the embryo.

Conjugation: (L. *cum*, together; *jugare*, to yoke.) Term used to describe mating between bacteria.

Contractualism: Moral theories which attempt to ground principles of morality in some hypothetical procedure of collective choice.

'Crippled' (virus): see Biological containment.

Cryopreservation: (Gr. *kryos*, cold.) The use of low temperatures to store living entities, e.g. organs, embryos or tissues.

Cytogenetics: The area of study that links the structure and behaviour of chromosomes with inheritance.

Cytoplasm: (Gr. *kytos*, hollow; *plasma*, mould.) The living contents of a cell excluding the nucleus.

Diffusion (of an innovation): The spread of an innovation, with or without modification, through a population of potential users.

DNA: Deoxyribonucleic acid; the molecule which for all organisms except RNA viruses encodes information for the reproduction and functioning of cells, and for the replication of the DNA molecule itself; information encoded in DNA molecules is transmitted from generation to generation.

DNA base pair: A pair of DNA nucleotide bases; one of the pair is on one chain of the duplex DNA molecule, the other is on the complementary chain; they pair across the double helix in a very specific way: adenine (A) can only pair with thymine (T); cytosine (C) can only pair with guanine (G); the specific nature of base pairing enables accurate replication of the chromosomes and helps to maintain the constant composition of the genetic material (see also Base; Nucleotide).

DNA fingerprinting: A technique which uses DNA probes to generate personal genetic profiles which are as specific to individuals as conventional fingerprints.

DNA probe: A short piece of DNA that is used to detect the presence of a complementary piece of DNA in a sample of DNA under analysis; used in genetic screening, for example.

DNA sequence: The order of base pairs in the DNA molecule; genetic information can be encoded in the sequence of bases.

DNA technology: See Microgenetic engineering.

Dominant genetic disorder: (L. *dominans*, ruling.) A disorder which is expressed in the phenotype when the gene responsible is inherited from just one parent (cf. Recessive genetic disorder).

Double helix: The name given to the structure of the DNA molecule; two complementary strands which lie alongside and twine around each other, joined by cross-linkages between base pairs (see DNA base pair).

Downstream process: A process in industrial biotechnology which occurs after the bioconversion stage; for example, product recovery, separation and purification.

Duplicative transposition: The process whereby transposable genetic elements move around genomes; a copy of a transposable genetic element located on a chromosome is duplicated and then deposited at a new location without loss of the original sequence.

EC Directive: EC Community law; binding on Member States as to ends but not means; can be issued by the European Commission or the Council of Ministers.

Ecology: Term coined in 1866 by Ernst Haeckel (1834–1919) to describe the branch of biology dealing with inter-relations between organisms and their environment.

Ecosystem: (Gr. *oikos,* home; L. *systema,* an assemblage of things adjusted into a regular whole.) A unit made up of all the living and non-living components of a particular area that interact and exchange materials with each other.

Elution: (L. *elutio,* wash.) Washing away impurity; cleanse.

Enablement requirement: A patentor's legal obligation to provide full technical details of his or her novel process or product which should allow a person with ordinary skill in the field to duplicate the invention.

Endocrine system: (Gr. *endon,* in; *krinein,* to separate.) Secretion of hormones from ductless glands (see Hormone).

Enzyme: (Gr. *en,* in; *zyme,* leaven.) A biological catalyst produced by living cells; a protein molecule which mediates and promotes a chemical process without itself being altered or destroyed; enzymes act with a given compound, the substrate, to produce a complex, which then forms the products of the reaction; enzymes are extremely efficient catalysts and very specific to particular reactions; the active principle of a ferment.

Epizootic: (Gr. *epi,* upon; *zoon,* animal.) Disease affecting a large number of animals simultaneously, corresponding to 'epidemic' in humans.

Escherichia coli (*E. coli*): A bacterial species that inhabits the intestinal tract of most vertebrates and on which much genetic work has been done; some strains are pathogenic to humans and other animals; many non-pathogenic strains are used experimentally as hosts for recombinant DNA.

Eugenics: (Gr. *eu,* well; *genos,* birth.) After Galton (1883), the science which deals with all the influences that improve the inborn qualities of a race; also with those that develop them to the utmost advantage.

Eukaryotes: Cells or organisms whose DNA is organised into chromosomes with a protein coat and sequestered in a well-defined cell nucleus; all living organisms except bacteria and blue-green algae are eukaryotic (cf. Prokaryotes).

Evolution (biological): (L. *evolvere,* to unroll.) Changes in DNA that occur during the history of organisms; the development of new organisms from pre-existing organisms since the beginning of life.

Expression (of genes): See Gene expression.

Factor VIII: One of approximately 13 substances – or factors – involved in blood-clotting; haemophilia A (classical haemophilia), a blood-clotting disorder, is caused by a deficiency of Factor VIII and is treated by administration of exogenous Factor VIII (see also Haemophilia).

Factor IX: A blood-clotting factor; exogenous Factor IX is used to treat Haemophilia B (Christmas disease), a blood-clotting disorder caused by a deficiency of endogenous Factor IX (see also Haemophilia).

Feedstock: The raw material used for the production of chemicals.

Fermentation: The anaerobic (without oxygen) biological conversion of organic molecules, usually carbohydrates, into alcohol, lactic acid and gases; it is brought about by enzymes either directly or as components of certain bacteria and in yeasts; in general use, the term is sometimes applied to bioprocesses which are not, strictly speaking, fermentation.

Fertilisation: In sexually reproducing organisms, the activation of the development of an egg through the union of sperm with the egg, so combining their genetic complements.

Gene: (Gr. *genos,* descent.) A gene is a section of a nucleic acid molecule in which the sequence of bases encodes the structure of, or is involved in the synthesis of, a protein.

Gene enhancement: The insertion of additional genetic material into an otherwise normal genome in order to enhance a trait perceived of as desirable; an example is the insertion of additional growth hormone genes into the genome of a normal individual in order to increase his or her height or rate of growth over what is considered to be the normal level.

Gene expression: The mechanism whereby the genetic instructions in a given cell are decoded and processed into the final functioning product, usually a protein.

Gene mapping: Determining the relative locations (loci) of genes on chromosomes.

Gene probe: A short piece of DNA or RNA used to detect the presence of complementary sequences in other nucleic acid molecules.

Gene sequencing: Determining the sequence of bases in a molecule of DNA.

Gene splicing: see *In vitro* genetic recombination.

Gene therapy: The correction of the effect of a genetic defect in an organism or cell by direct intervention with the genetic material; one method under investigation is gene replacement therapy in which additional foreign DNA is inserted to compensate for the malfunctioning gene; another approach would be to activate dormant genes within the genome whose function would substitute for the missing function of the malfunctioning gene.

Genetic code: The relationship between the sequence of bases in the nucleic acids of genes and the sequence of amino acids in the proteins that they code for.

Genetic determinism: The theory that the phenotype is an innate and essentially unchangeable expression of the genotype.

Genetic disorder: A disorder which is associated with a specific defect in the hereditary material; may or may not be congenital (for example, late-onset genetic disorders are not manifested at birth) and may or may not be inherited (for example, chromosomal abnormalities which are newly arisen).

Genetic enfeeblement: see Biological containment.

Genetic engineering: The manipulation of heredity or the hereditary material; the direct and deliberate attempts by humans to influence the course of evolution and to alter its products; the basic techniques are mutagenesis and hybridisation, which introduce genetic variation, and artificial selection which biases quantitatively the genetic variation of subsequent generations; genetic engineering using artificial selection and traditional hybridisation through cross-breeding has been long practised; artificial mutagenesis is a twentieth century development; in the second half of the twentieth century all three techniques have been developed by *in vitro* methods including microgenetic engineering and cell fusion.

Genetic fingerprinting: see DNA fingerprinting.

Genetic manipulation: see Microgenetic engineering.

Genetic marker: An identifiable feature encoded in the genetic material of an organism; an example of the use of genetic markers is the insertion of a small unique and inactive piece of DNA into the genome of organisms to be released into the environment to enable such organisms to be identified upon their recovery from the environment.

Genetic material: DNA, genes and chromosomes which constitute an organism's hereditary material; RNA in certain viruses.

Genetic recombination: The excision and rejoining of DNA molecules; formation of a new association of genes or DNA sequences from different parental origins (see also *In vitro* genetic recombination and Recombinant DNA techniques).

Genetic screening: A range of techniques used to diagnose phenotypic traits which have or are believed to have a genetic basis; largely used to detect such traits before they become evident, but can also be used to verify the diagnosis of traits after they have become apparent.

Genetically 'crippled': see Biological containment.

Genetics: (L. *genesis*, origin, descent.) That part of biology dealing with both the constancy of inheritance and its variation; the study of the replication, transmission and expression of hereditary information.

Genome: A collective noun for all the genetic information that is typical of a particular organism; every somatic cell in a multicellular organism contains a full genome; the term genome is also applied to the genetic contents characteristic of major groups (for example, the eukaryotic genome) or of a species (for example, the human genome); not all portions of a genome are genes (i.e. genomes include non-coding DNA); genomes do not include the genetic material of extrachromosomal elements, nor of plasmids or viruses harboured by a cell, although this distinction between genomic and non-genomic genetic material is a reflection of a static paradigm of the genome which is increasingly believed to be inaccurate; genomes can be regarded as an ecosystem of genetic elements (see Wheale and McNally (1988), particularly Chapter 4).

Genotype: (Gr. *genos*, race; *typos*, image.) The genetic constitution of an organism with respect to a particular genetic trait, for example, eye colour (cf. Phenotype).

Germ: A popular word for a microorganism; also used in the eighteenth and nineteenth centuries to describe the hereditary material.

Germ cell: Sex cell or a cell which gives rise to sex cells.

Germline: Cells from which sex cells are derived.

Germplasm: The term for that part of an organism which passed on hereditary characteristics to the next generation. According to Weismann's germplasm theory of 1883, the germplasm was transmitted from generation to generation in germ cells.

Green Revolution: Term used to describe the replacement of traditional crops by high-yield varieties requiring irrigation systems and inputs of fertilisers and pesticides to sustain them.

Growth hormone: In animals, a hormone which affects a large number of metabolic processes including the regulation of growth; somatotropin (or somatotrophin) is a growth hormone produced by the anterior lobe of the pituitary gland (see Hormone).

Haemoglobin: (Gr. *haima*, blood; L. *globus*, a ball.) The red-coloured protein which binds and carries oxygen in red blood cells.

Haemophilia: (Gr. *haima*, blood; *philos*, inclined to.) Constitutional tendency to haemorrhage as a result of abnormal blood clotting; inherited as a sex-linked recessive single-gene disorder; results in pain, profound anaemia and orthopaedic problems caused by bleeding into the joints; caused by a deficiency of one of the clotting factors of the blood factor VIII in the case of haemophilia A (classical haemophilia), and factor IX in the case of haemophilia B (Christmas disease) (see also Factor VIII and Factor IX).

Hereditary disease: A disorder of the genetic material which is transmissible from generation to generation.

Heterozygote: (Gr. *heteros*, other; *zygon*, yoke) an organism or cell having two different alleles at corresponding loci on homologous chromosomes.

Hormone: A chemical messenger of the body carried in the bloodstream from the gland which secretes it to a target organ where it has a regulatory effect; an example is insulin.

Host: A cell (microbial, animal or plant) whose metabolism is used for the reproduction of a virus, plasmid or other form of foreign DNA, including vectors and recombinant DNA.

Hybrid: (L. *hybrida*, cross.) A molecule, cell or organism produced by combining the genetic material of genetically dissimilar organisms; traditionally, hybrids were produced by interbreeding whole animals or plants; cell fusion technology and transgenic manipulation are innovations in hybridisation.

Hybridoma: A 'hybrid myeloma'; a cell produced by the fusion of an antibody-producing cell (lymphocyte) with a cancer cell (myeloma).

Hybridoma technology: The technology of fusing antibody-producing cells with tumour cells to produce in hybridomas which proliferate continuously and produce monoclonal antibodies.

Immune system: The body's system of defence against invasion by foreign organisms and certain chemicals.

Immunogenic (L. *immunis*, free; *genos*, birth) causing formation of antibodies.

In utero: (L. *uterus*.) In the womb.

In vitro: (L. *vitrum*, glass.) Literally in glass; biological processes studied and manipulated outside of the living organism.

***In vitro* fertilisation** (IVF): The fertilisation of an egg cell by sperm on a glass dish.

***In vitro* genetic recombination**: The precise excision and joining of DNA fragments on the laboratory bench exploiting the biochemical tools of the cell – restriction enzymes and DNA ligase – and the inherent pairing affinity of the duplex DNA molecule (see Recombinant DNA techniques and Genetic recombination).

In vivo: (L. *vivo*, live.) Within the living organism.

Infection: The invasion of an organism, or part of an organism, by pathogenic microorganisms, for example, viruses or bacteria.

Innovation: The first introduction of a new product, process or system into the ordinary commercial or social activity.

Insertional mutagenesis: In transgenic manipulation, the mutation of target host cell genes by the integration of foreign genes.

Insulin: A hormone, the release of which lowers the level of glucose sugar in the blood.

Interferons: A class of proteins released by certain mammalian cells in response to various stimuli, including viral infection, which are thought to inhibit viral replication; undergoing clinical trials as anti-viral and anti-cancer agents.

Intron: A nucleic acid sequence within a gene which is transcribed into RNA but then excised from the RNA transcript before it is translated into protein.

Invention: The first idea, sketch or contrivance of a new product, process or system.

Invertebrates: (L. *in*, not; *vertebra*, joint.) A general term for all animal groups except the vertebrates, i.e., all animals without a backbone; includes insects, worms and crustaceans.

Jumping genes: See Mobile genetic elements.

Karyotype: The chromosomal constitution of an individual.

Ketosis: A condition, found in fasting animals and diabetic animals, in which large amounts of ketone bodies appear in the blood (ketonaemia) and urine (ketonuria) as a result of the breakdown of body fat.

Laparoscopy: Examination of abdominal structures using a laparascope (an illuminated tubular instrument) which is passed through a small incision in the wall of the abdomen. Its uses include: diagnosis; biopsy; minor pelvic surgery; and collecting ova (eggs) for *in vitro* surgery.

Lesch–Nyhan syndrome: An X-linked recessive single-gene disorder in which the central nervous system gradually deteriorates; affected children are typically spastic and slow-growing and under compulsion to bite their lips and fingers so that their arms must be restrained to prevent self-mutilation; the biochemical basis is a deficiency of the enzyme hypoxanthine-guanine phosphoribosyl transferase (HPRT).

Ligase: An enzyme which catalyses the joining together of two molecules; for example, DNA ligase catalyses the joining of two DNA molecules.

Mendelian: After Gregor Mendel (1822–84), relating to the Mendelian theory of heredity; the pattern of inheritance exhibited by traits controlled by a single gene in which there is a simple dominant or recessive relationship between alleles.

Microbe: An alternative term for a microorganism (see Microorganism).

Microbiology: The study of microorganisms.

Microgenetic engineering: Following Wheale and McNally (1988), the techniques which enable the molecular biologist to decode, compare, construct, mutate, excise, join, transfer and clone specific sequences of DNA, thus directly manipulating the genetic material to produce organisms, cells and subcellular components; applications include scientific research, biotechnology, farming, health care and biological defence (see also *In vitro* genetic recombination and Recombinant DNA techniques).

Microorganism: An organism belonging to the categories of viruses, bacteria, fungi, algae or protozoa; microorganisms or 'animalcules' as they were called when first observed by Anton van Leeuwenhoek (1632–1723) of Delft in Holland in the seventeenth century; until the second half of the nineteenth century, when the nature of their association with putrefaction, fermentation and disease became a focus of microscopy, Leeuwenhoek's discovery of parasites and bacteria remained a 'curiosity', and those who continued to search for *Vermes chaos,* (see also *Vermes chaos*) as Linnaeus classified these 'incredibly small animals', were regarded as eccentrics.

Mobile genetic elements: RNA and DNA sequences that move from one place to another both within and between genomes; popularly known as 'jumping genes'; the family of mobile genetic elements includes viruses, subviral infectious elements, plasmids, RNA introns, messenger RNA molecules, transposable genetic elements and oncogenes; characteristic features include the ability to move pieces of DNA either within or between cellular genomes, the ability to usurp the biochemistry of the host cell to bring about their own replication, the ability to alter the structure of the host cell genome, and the ability to modify the expression of other genetic elements within the host cell; they are believed to play a crucial role in evolution, and are implicated in development and pathology; viruses and plasmids are mobile genetic elements which are used as vectors in gene transfer technology.

Molecule: A group of two or more atoms joined together by chemical bonds.

Monoclonal antibody: One of a clone of antibodies produced by a hybridoma.

Multifactorial disorder: A disorder in which both genetic and exogenous factors are multiple and interact; the genetic part of multifactorial causation is polygenic (the result of the action of many genes).

Mutagen: (L. *mutare,* to change; Gr. *gennaein,* to generate.) A chemical or physical or other agent which increases the frequency of mutation.

Mutant: An organism or gene which deviates by mutation from the parent organism(s) or gene in one or more characteristics.

Mutation: A change in the genetic material; can refer to changes in a single DNA base pair or in a single gene, and also to changes in chromosome structure and number which are recognisable under the microscope; mutation in the germline or sex cells could result in genetic illness or changes of evolutionary significance; somatic cell mutation may be the basis of some cancers and some aspects of ageing.

Neo-Luddite: Person actively opposed to the introduction of new technology; after the Luddites, an organised band of mechanics who went about destroying machinery in the Midlands and North of England in the early nineteenth century; derived from Ned Lud, who, in the latter half of the eighteenth century, smashed up machinery belonging to a Leicestershire manufacturer as a protest against mechanisation.

Newcastle disease: Avian influenza.

Nuclear polyhedrosis virus (NPV): Subgroup of the baculoviruses; their name derives from a gene they have which codes for a protein called the polyhedrin protein; during the dispersive phase of their life cycle, many NPVs are encased together in a polyhedral structure comprised of polyhedrin which protects them from adverse environmental conditions and increases their environmental persistence (see also Baculovirus; Polyhedrin inclusion body).

Nucleic acids: Either RNA or DNA; complex organic molecules composed of sequences of nucleotides.

Nucleotide: (L. *nucleus*, kernel.) In DNA, a molecular grouping comprised of a base plus a sugar molecule (deoxyribose) plus a phosphate group; DNA is a polynucleotide, that is a molecule comprised of many nucleotides (see Base).

Nucleus: A region in the cells of eukaryotes, surrounded by a membrane, in which the main chromosomes are sequestered.

Oncogene: (Gr. *onchos*, swelling; L. *genesis*, origin.) Found in every cell of the body, it is postulated that oncogenes are a broad class of regulatory genes that control the activity of other genes; the oncogene theory of cancer is that when an oncogene is activated at an inappropriate stage in the life cycle of an individual, the cell begins to multiply in an uncontrolled way; cellular oncogenes may be derived from viruses which have integrated into the genome of a host cell, or conversely, certain viruses may have acquired cellular oncogenes when they became infectious.

Oncogenic: cancer-causing.

Oncomice: also known as 'cancer mice'; genetically engineered mice with human cancer genes (or oncogenes: see Oncogene) inserted into the genome of mice to cause the development of breast cancer for the purpose of scientific study of how breast cancer develops and to test new drugs and therapies that may be useful for the treatment of cancer in humans.

Painience: the capacity of an organism to suffer pain.

Patent: The exclusive right to a property in an invention; this monopoly on invention gives its owner the legal right of action against anyone exploiting the patented research without the patentor's consent.

Pathogen: (Gr. *pathos,* suffering.) An organism which causes disease.
Pesticides: Herbicides, insecticides and fungicides used in farming and forestry to control the organisms which reduce the quality or quantity of crop yield.
Phage: Abbreviation of bacteriophage, a virus which infects bacteria.
Phenotype: (Gr. *phainein,* to show; *typos,* image.) The manifest expression of the genetic determinants for a particular trait – the genotype (cf. Genotype).
Physical containment: Measures that are designed to prevent or minimise the escape of recombinant organisms.
Plaque: A zone of clearing on the otherwise opaque areas of dense bacterial growth formed when bacteria which have been infected with viruses are grown on nutrient agar; each plaque is the derived from a single virus; the plaque assay method exploits this phenomenon to measure the number of bacterial viruses (bacteriophages) suspended in a given volume of medium.
Plasmid: A small circle of DNA usually found in prokaryotes, which replicates independently of the main chromosome(s) and can be transferred naturally from one organism to another, even across species boundaries; plasmids and some viruses are used as vectors in transgenic manipulation.
Polypeptide: Long folded chains of amino acids; proteins are made from polypeptides.
Prokaryotes: Organisms whose genetic material is not sequestered in a well-defined nucleus; includes bacteria and blue-green algae (cf. Eukaryotes).
Promoter: In transcription, a DNA sequence to which the enzymes which catalyse messenger RNA synthesis bind.
Pronucleus: (L. *pro,* before; *nucleus,* kernel.) Term used for the nucleus of a mature egg cell or mature sperm.
Prophylactic: (Gr. *prophylaktikos.*) Preventing disease.
Protein: (Gr. *proteion,* first.) Proteins are polypeptides, i.e. they are made up of amino acids joined together by peptide links; acting as hormones, enzymes and connective and contractile structures, proteins endow cells and organisms with their characteristic properties of shape, metabolic potential, colour and physical capacities.
Proteolysis: (Gr. *proteion,* first; *lysis,* loosing.) In digestion, the breaking down of dietary proteins into their constituent amino acids by enzymes in the gastrointestinal tract.
Protoplast: Plant cell whose cell wall has been deliberately removed; used in plant cell fusion.
Recessive genetic disorder: (L. *recessus,* withdrawn.) The gene mutations responsible for recessive disorders are usually only harmful to individuals who do not have a corresponding normal gene; except in the case of sex-linked conditions, a recessive disorder is only expressed when the gene mutation responsible

is inherited from both parents (see Dominant genetic disorder).

Recombinant bacterium/cell/plasmid/vector/virus etc.: Contains recombinant DNA (or recombinant RNA).

Recombinant DNA: A hybrid DNA molecule which contains DNA from two distinct sources (see Genetic recombination).

Recombinant DNA techniques (technology): A type of microgenetic engineering; the combination of *in vitro* genetic recombination techniques with techniques for the insertion, replication and expression of recombinant DNA inside living cells.

Restriction enzymes: Bacterial enzymes that cut DNA at specific DNA sequences; exploited by the microgenetic engineer, for example, in the execution of the precise excision of DNA fragments for *in vitro* genetic recombination, and in the analysis of differences between DNA molecules.

Retroviruses: Viruses which encode their hereditary information in the nucleic acid RNA; retroviruses are being engineered for use as vectors in human gene therapy.

Reverse transcription: Synthesis of a single strand of DNA using an RNA template.

Ribo(se)nucleic acid (RNA): A molecule which resembles DNA in structure; acts as an adjunct in the execution and mediation of the genetic instructions encoded in DNA, for example, messenger RNA (mRNA) is transcribed from a single strand of a DNA molecule and is the template on which amino acids align to form the coded protein; RNA has been observed to function as an autocatalyst in transcription and possibly controls translation; RNA may have been the primordial genetic molecule functioning both as an enzyme and as a self-replicating repository of genetic information; RNA is the repository of hereditary information for some viruses (retroviruses).

Scale-up: The transition of a process from laboratory scale to industrial scale.

Scientism: The belief that the scientific approach is objective and the only rational way to approach any problem; it is argued that this belief promotes a passive acceptance of techniques and technologies which are scientifically based and thus apolitical.

Sentient: (L. *sentiens.*) Having the faculty of perception; a sentient being is one who perceives; refers to all animals which have the capacity for conscious experiences such as pleasure or pain.

Sentientism: The moral position that holds that it is wrong to cause pain or distress to any sentient being, unless it is with their agreement, or unless it will bring unquestionable benefit to that same individual sentient if he or she is unable to give informed consent.

Sex cells: Cells which fuse together to form a fertilised egg; in human beings the male sex cell is the sperm, and the female sex cell is the egg cell, also known as the ovum (pl. ova) (cf. Somatic cell).

Sex chromosomes: Chromosomes that differ in number or morphology in different sexes and contain genes determining sex type; in human beings the sex chromosomal constitution of a normal female is XX and that of a normal male is XY.

Sex-linked (X-linked) trait: Determined by a gene located on a sex chromosome, usually the X chromosome in humans, and hence the term 'X-linked trait' is sometimes preferred.

Shotgunning (in microgenetic engineering): A technique for breaking up the entire genome of an organism into small pieces and then inserting those pieces into a host cells, where they are cloned into a gene library for the organism.

Somatic cell: (Gr. *soma*, body.) Any cell of the body other than germ cells (cf. Germ cell).

Somatotropin/somatotrophin: (Gr. *soma*, body; *trephein*, to increase.) A protein hormone which is produced by the anterior lobe of the pituitary gland and promotes general body growth.

Speciation: The origin of a new species.

Species: (L. *species*, particular kind.) A unit of biological classification; sexually reproducing organisms are classified as belonging to the same species if they can interbreed and produce fertile offspring; the interpretation of what constitutes a species is controversial; it is especially difficult to apply the concept of species to bacteria which are not subject to reproductive isolation.

Taxonomy: (Gr. *taxis*, arrangement; *nomos*, law.) The method of arrangement or classifying, particularly of living organisms. Taxonomic studies have led to the development of a system of classification which divides all living things into two large groups called kingdoms, each of which is divided into a series of major subgroups called phyla (sing. phylum). Each phylum is further divided into a series of successively smaller groups known as classes, orders, families, genera (sing. genus) and finally species. There is generally only one kind of organism in a species. In modern taxonomy, an organism is named using the binomial system, under which an organism's name is designated by the genus to which it belongs (generic name), followed by the name of the species (specific name). The name is always written with a capital letter and the specific name with a small letter, for example, the taxonomic name of humans is *Homo sapiens*. The Swedish naturalist Carl Linnaeus introduced the binomial system of naming organisms in 1735.

Telos: (Gr. end.) The nature or purpose of a thing or creature.

Teratogeny: (Gr. *teras*, monster; *genos*, birth.) The formation of monstrous foetuses or births; thalidomide was a classified as a teratogenic drug.

Test-tube babies: Popular term for the resultant offspring of successful *in vitro* fertilisation, implantation, pregnancy and birth.

Tissue plasminogen activator (TPA): Biological drug for dissolving blood clots; may have therapeutic value in the treatment of heart attack patients.

Toxin: A substance, in some cases produced by disease-causing microorganisms, that is poisonous to living organisms.

Transcription: (L. *transcribere*, to copy out.) In gene expression, the synthesis of an RNA molecule using a section of one strand of a DNA molecule as a template.

Transduction: (L. *transducere*, to transfer.) In genetics, the transfer of genetic material between cells mediated by an infectious mobile genetic element for example, a virus.

Transformation: (L. *transformare*, to change in shape.) In genetics, the process whereby a piece of foreign 'naked' DNA is taken up from the surrounding medium by a cell into which it integrates giving that cell new properties.

Transgenic manipulation: A type of microgenetic engineering in which genetic material from one species is inserted into the genome of a different species; an application of recombinant DNA technology.

Transgenic organism: An organism which has been microgenetically engineered so that its genome contains genetic material derived from a different species.

Translation: (L. *translatio*, transferring.) In protein synthesis, the conversion of the base sequence of a messenger RNA molecule (mRNA) into a polypeptide chain of amino acids for the construction of a particular protein.

Translocation (chromosomal): (L. *trans*, across; *locus*, place.) In genetics, the displacement of part or all of one chromosome to another.

Transposable genetic element: A genetic element that moves from site to site within a cellular genome; now believed to be a major feature of all DNA (see Mobile genetic element).

Trypanosome: Parasitic microorganism which causes diseases, for example, sleeping sickness, in humans and other animals.

Utilitarianism: The philosophy that claims the ultimate good to be the greatest happiness of the greatest number and defines the rightness of actions in terms of their contribution to the general happiness; it follows that no specific moral principle is absolutely certain and necessary, since the relation between actions and their consequences varies with the circumstances.

Vaccine: A substance introduced into an animals's body in order to stimulate the activation of the body's immune system as a precautionary measure against future exposures to a particular pathogenic agent.

Vectors: Vectors are self-replicating entities used as vehicles to transfer foreign genes into living cells and then replicate and possibly also express them; examples are plasmids and viruses.

Vermes chaos: L. *vermis*, a worm; *vermin*, a noxious animal; animals destructive to game; animals injurious to crops and other possessions; noxious persons, in contempt. Gr. *chaos*, that confusion in which matter was supposed to have existed before it was reduced to order; confusion; disorder (see Microorganism).

Vertebrate: (L. *vertebrae*, a joint.) Animal with a backbone; fish, amphibians, reptiles, birds and mammals are vertebrates.

Viroid: Pieces of 'naked' (without a protein 'coat') infectious RNA.

Virus: A minute infectious agent; a mobile genetic element; composed of nucleic acid (DNA or RNA) wrapped in a protein coat; can survive on its own but in order to replicate it must be inside a living cell; viruses are used as vectors in microgenetic engineering.

Xenografting: (Gr. *xenos*, a stranger.) The transplantation of an organ or tissue between different species; also known as heterograft Gr. *heteros*, another (see also Allografting).

Name Index

Subject Index